D1318564

Business Analytics Principles, Concepts, and Applications with SAS

What, Why, and How

Marc J. Schniederjans

Dara G. Schniederjans

Christopher M. Starkey

Associate Publisher: Amy Neidlinger
Executive Editor: Jeanne Glasser Levine
Operations Specialist: Jodi Kemper
Cover Designer: Alan Clements
Cover Image: Alan McHugh
Managing Editor: Kristy Hart
Senior Project Editor: Betsy Gratner
Copy Editor: Gill Editorial Services
Proofreader: Chuck Hutchinson
Indexer: Erika Millen
Senior Compositor: Gloria Schurick
Manufacturing Buyer: Dan Uhrig

© 2015 by Marc J. Schniederjans, Dara G. Schniederjans, and Christopher M. Starkey
Published by Pearson Education, Inc.
Upper Saddle River, New Jersey 07458

For information about buying this title in bulk quantities, or for special sales opportunities (which may include electronic versions; custom cover designs; and content particular to your business, training goals, marketing focus, or branding interests), please contact our corporate sales department at corpsales@pearsoned.com or (800) 382-3419.

For government sales inquiries, please contact governmentsales@pearsoned.com.

For questions about sales outside the U.S., please contact international@pearsoned.com.

Company and product names mentioned herein are the trademarks or registered trademarks of their respective owners.

Printed in the United States of America

First Printing: October 2014

ISBN-10: 0-13-398940-2
ISBN-13: 978-0-13-398940-3

Pearson Education LTD.
Pearson Education Australia PTY, Limited.
Pearson Education Singapore, Pte. Ltd.
Pearson Education Asia, Ltd.
Pearson Education Canada, Ltd.
Pearson Educación de Mexico, S.A. de C.V.
Pearson Education—Japan
Pearson Education Malaysia, Pte. Ltd.

Library of Congress Control Number: 2014945193

This book is dedicated to Miles Starkey.
He is what brings purpose to our lives
and gives us a future.

Contents-at-a-Glance

Table of Contents

About the Authors

Marc J. Schniederjans is the C. Wheaton Battey Distinguished Professor of Business in the College of Business Administration at the University of Nebraska-Lincoln and has served on the faculty of three other universities. Professor Schniederjans is a Fellow of the Decision Sciences Institute (DSI) and in 2014–2015 will serve as DSI's president. His prior experience includes owning and operating his own truck leasing business. He is currently a member of the Institute of Supply Management (ISM), the Production and Operations Management Society (POMS), and Decision Sciences Institute (DSI). Professor Schniederjans has taught extensively in operations management and management science. He has won numerous teaching awards and is an honorary member of the Golden Key honor society and the Alpha Kappa Psi business honor society. He has published more than a hundred journal articles and has authored or coauthored twenty books in the field of management. The title of his most recent book is *Reinventing the Supply Chain Life Cycle*, and his research has encompassed a wide range of operations management and decision science topics. He has also presented more than one hundred research papers at academic meetings. Professor Schniederjans is serving on five journal editorial review boards, including *Computers & Operations Research, International Journal of Information & Decision Sciences, International Journal of Information Systems in the Service Sector, Journal of Operations Management*, and *Production and Operations Management*. He is also serving as an area editor for the journal *Operations Management Research* and as an associate editor for the *International Journal of Strategic Decision Sciences* and *International Journal of the Society Systems Science and Management Review: An International Journal* (Korea). In addition, Professor Schniederjans has served as a consultant and trainer to various business and government agencies.

Dara G. Schniederjans is an assistant professor of Supply Chain Management at the University of Rhode Island, College of Business Administration. She has published articles in journals such as *Decision Support Systems*, *Journal of the Operational Research Society*, and *Business Process Management Journal*. She has also coauthored two text books and coedited a readings book. She has contributed chapters to readings utilizing quantitative and statistical methods. Dara has served as a guest coeditor for a special issue on *Business Ethics in Social Sciences* in the *International Journal of Society Systems Science*. She has also served as a website coordinator for Decisions Sciences Institute. She currently teaches courses in Supplier Relationship Management and Operations Management.

Christopher M. Starkey is an economics student at the University of Connecticut-Storrs. He has presented papers at the Academy of Management and Production and Operations Management Society meetings. He currently teaches courses in Principles of Microeconomics and has taught Principles of Macroeconomics. His current research interests include macroeconomic and monetary policy, as well as other decision-making methodologies.

Preface

Like the face on the cover of this book, we are bombarded by information every day. We do our best to sort out and use the information to help us get by, but sometimes we are overwhelmed by the abundance of data. This can lead us to draw wrong conclusions and make bad decisions. When you are a global firm collecting millions of transactions and customer behavior data from all over the world, the size of the data alone can make the task of finding useful information about customers almost impossible. For that firm and even smaller businesses, the solution is to apply *business analytics* (BA). BA helps sort out large data files (called "big data"), find patterns of behavior useful in predicting the future, and allocate resources to optimize decision-making. BA involves a step-wise process that aids firms in managing big data in a systematic procedure to glean useful information, which can solve problems and pinpoint opportunities for enhanced business performance.

This book has been written to provide a basic education in BA that can serve both academic and practitioner markets. In addition to bringing BA up-to-date with literature and research, this book explains the BA process in simple terms and supporting methodologies useful in its application. Collectively, the statistical and quantitative tools presented in this book do not need substantial prerequisites other than basic high school algebra. To support both markets, a substantial number of solved problems are presented along with some case study applications to train readers in the use of common BA tools and software. Practitioners will find the treatment of BA methodologies useful review topics. Academic users will find chapter objectives and discussion questions helpful for serving their needs while also having an opportunity to obtain an Instructor's Guide with chapter-end problem solutions and exam questions.

The purpose of this book is to explain what BA is, why it is important to know, and how to do it. To achieve this purpose, the book presents conceptual content, software familiarity, and some analytic tools.

Conceptual Content

The conceptual material is presented in the first eight chapters of the book. (See Section 1.4 in Chapter 1 for an explanation of the book's organization.) The conceptual content covers much more than what BA is about. It explains why BA is important in terms of providing answers to questions, how it can be used to achieve competitive advantage, and how to align an organization to make best use of it. The book explains the managerial aspects of creating a BA presence in an organization and the skills BA personnel are expected to possess. The book also describes data management issues such as data collection, outsourcing, data quality, and change management as they relate to BA.

Having created a managerial foundation explaining "what" and "why" BA is important, the remaining chapters focus on "how" to do it. Embodied in a three-step process, BA is explained to have descriptive, predictive, and prescriptive analytic steps. For each of these steps, this book presents a series of strategies and best practice guides to aid in the BA process.

Software

Much of what BA is about involves the use of software. Unfortunately, no single software covers all aspects of BA. Many institutions prefer one type of software over others. To provide flexibility, this book's use of software provides some options and can be used by readers who are not even interested in running computer software. In this book, SAS® and Lingo® software are utilized to model and solve problems. The software treatment is mainly the output of these software systems, although some input and instructions on their use are provided. For those not interested in running software applications, the exposure to the printouts provides insight into their informational value. This book recognizes that academic curriculums prefer to uniquely train students in the use of software and does not duplicate basic software usage. As a prerequisite to using this book, it is recommended that those interested in running software applications for BA become familiar with and are instructed on the use of whatever software is desired.

Analytic Tools

The analytic tool materials are chiefly contained in this book's appendixes. BA is a statistical, management information system (MIS) and quantitative methods tools-oriented subject. Although the conceptual content in the book overviews how to undertake the BA process, the implementation of how to actually do BA requires quantitative tools. Because some practitioners and academic programs are less interested in the technical aspects of BA, the bulk of the quantitative material is presented in the appendixes. These appendixes provide an explanation and illustration of a substantial body of BA tools to support a variety of analyses. Some of the statistical tools that are explained and illustrated in this book include statistical counting (permutations, combinations, repetitions), probability concepts (approaches to probability, rules of addition, rules of multiplication, Bayes's theorem), probability distributions (binomial, Poisson, normal, exponential), confidence intervals, sampling methods, simple and multiple regression, charting, and hypothesis testing. Although management information systems are beyond the scope of this book, the software applications previously mentioned are utilized to illustrate search, clustering, and typical data mining applications of MIS technology. In addition, quantitative methods and tools explained and illustrated in this book include linear programming, duality and sensitivity analysis, integer programming, zero-one programming, forecasting modeling, nonlinear optimization, simulation analysis, breakeven analysis, and decision theory (certainty, risk, uncertainty, expected value opportunity loss analysis, expected value of perfect information, expected value of imperfect information).

We want to acknowledge the help of individuals who provided needed support for the creation of this book. First, we really appreciate the support of our editor, Jeanne Glasser Levine, and the outstanding staff at Pearson. They made creating this book a pleasure and worked with us to improve the final product. Decades of writing books with other publishers permitted us to recognize how using a top-tier publisher like we did makes a difference. We thank Alan McHugh, who developed the image on our book cover. His constant willingness to explore and be innovative with ideas made a significant contribution to our book. We also want to acknowledge the great editing help we received from Jill Schniederjans. Her skill has reduced the wordiness and enhanced the content (making parts less boring to read). Finally, we would like to acknowledge the help of Miles Starkey, whose presence and charm have lifted our spirits and kept us on track to meet completion deadlines.

Although many people have assisted in preparing this book, its accuracy and completeness are our responsibility. For all errors that this book may contain, we apologize in advance.

Marc J. Schniederjans

Dara G. Schniederjans

Christopher M. Starkey

Part I

What Is Business Analytics?

1

What Is Business Analytics?

Chapter objectives:

- Define business analytics.
- Explain the relationship of analytics and business intelligence to the subject of business analytics.
- Describe the three steps of the business analytics process.
- Describe four data classification measurement scales.
- Explain the relationship of the business analytics process with the organization decision-making process.

1.1 Terminology

Business analytics begins with a *data set* (a simple collection of data or a data file) or commonly with a *database* (a collection of data files that contain information on people, locations, and so on). As databases grow, they need to be stored somewhere. Technologies such as *computer clouds* (hardware and software used for data remote storage, retrieval, and computational functions) and *data warehousing* (a collection of databases used for reporting and data analysis) store data. Database storage areas have become so large that a new term was devised to describe them. *Big data* describes the collection of data sets that are so large and complex that software systems are hardly able to process them (Isson and Harriott, 2013, pp. 57–61). Isson and Harriott (2013, p. 61) define *little data* as anything that is not big data. Little data describes the smaller data segments or files that help individual businesses keep track of customers. As a means of sorting through data to find useful information, the application of analytics has found new purpose.

Three terms in business literature are often related to one another: analytics, business analytics, and business intelligence. *Analytics* can be defined as a process that involves the use of statistical techniques (measures of central tendency, graphs, and so on), information system software (data mining, sorting routines), and operations research methodologies (linear programming) to explore, visualize, discover, and communicate patterns or trends in data. Simply, analytics converts data into useful information. Analytics is an older term commonly applied to all disciplines, not just business. A typical example of the use of analytics is the weather measurements collected and converted into statistics, which in turn predict weather patterns.

There are many types of analytics, and there is a need to organize these types to understand their uses. We will adopt the three categories (*descriptive, predictive,* and *prescriptive*) that the *Institute of Operations Research and Management Sciences* (INFORMS) organization (www.informs.org) suggests for grouping the types of analytics (see Table 1.1). These types of analytics can be viewed independently. For example, some firms may only use descriptive analytics to provide information on decisions they face. Others may use a combination of analytic types to glean insightful information needed to plan and make decisions.

Table 1.1 Types of Analytics

Type of Analytics	Definition
Descriptive	The application of simple statistical techniques that describe what is contained in a data set or database. Example: An age bar chart is used to depict retail shoppers for a department store that wants to target advertising to customers by age.
Predictive	An application of advanced statistical, information software, or operations research methods to identify predictive variables and build predictive models to identify trends and relationships not readily observed in a descriptive analysis. Example: Multiple regression is used to show the relationship (or lack of relationship) between age, weight, and exercise on diet food sales. Knowing that relationships exist helps explain why one set of independent variables influences dependent variables such as business performance.
Prescriptive	An application of decision science, management science, and operations research methodologies (applied mathematical techniques) to make best use of allocable resources. Example: A department store has a limited advertising budget to target customers. Linear programming models can be used to optimally allocate the budget to various advertising media.

The purposes and methodologies used for each of the three types of analytics differ, as can be seen in Table 1.2. These differences distinguish *analytics* from *business analytics*. Whereas analytics is focused on generating insightful information from data sources, business analytics goes the extra step to leverage analytics to create an improvement in measurable business performance. Whereas the process of analytics can involve any one of the three types of analytics, the major components of business analytics include all three used in combination to generate new, unique, and valuable information that can aid business organization decision-making. In addition, the three types of analytics are applied sequentially (descriptive, then predictive, then prescriptive). Therefore, *business analytics* (BA) can be defined as a process beginning with business-related data collection and consisting of sequential application of descriptive, predictive, and prescriptive major analytic components, the outcome of which supports and demonstrates business decision-making and organizational performance. Stubbs (2011, p. 11) believes that BA goes beyond plain analytics, requiring a clear relevancy to business, a resulting insight that will be implementable, and performance and value measurement to ensure a successful business result.

Table 1.2 Analytic Purposes and Tools

Type of Analytics	Purpose	Examples of Methodologies
Descriptive	To identify possible trends in large data sets or databases. The purpose is to get a rough picture of what generally the data looks like and what criteria might have potential for identifying trends or future business behavior.	Descriptive statistics, including measures of central tendency (mean, median, mode), measures of dispersion (standard deviation), charts, graphs, sorting methods, frequency distributions, probability distributions, and sampling methods.
Predictive	To build predictive models designed to identify and predict future trends.	Statistical methods like multiple regression and ANOVA. Information system methods like data mining and sorting. Operations research methods like forecasting models.
Prescriptive	To allocate resources optimally to take advantage of predicted trends or future opportunities.	Operations research methodologies like linear programming and decision theory.

Business intelligence (BI) can be defined as a set of processes and technologies that convert data into meaningful and useful information for business purposes. Although some believe that BI is a broad subject that encompasses analytics, business analytics, and information systems (Bartlett, 2013, p.4), others believe it is mainly focused on collecting, storing, and exploring large database organizations for information useful to decision-making and planning (Negash, 2004). One function that is generally accepted as a major component of BI involves storing an organization's data in computer cloud storage or in data warehouses. Data warehousing is not an analytics or business analytics function, although the data can be used for analysis. In application,

BI is focused on querying and reporting, but it can include reported information from a BA analysis. BI seeks to answer questions such as what is happening now and where, and also what business actions are needed based on prior experience. BA, on the other hand, can answer questions like why something is happening, what new trends may exist, what will happen next, and what is the best course for the future.

In summary, BA includes the same procedures as plain analytics but has the additional requirement that the outcome of the analytic analysis must make a measurable impact on business performance. BA includes reporting results like BI but seeks to explain why the results occur based on the analysis rather than just reporting and storing the results, as is the case with BI. Analytics, BA, and BI will be mentioned throughout this book. A review of characteristics to help differentiate these terms is presented in Table 1.3.

Table 1.3 Characteristics of Analytics, Business Analytics, and Business Intelligence

Characteristics	Analytics	Business Analytics (BA)	Business Intelligence (BI)
Business performance planning role	What is happening, and what will be happening?	What is happening now, what will be happening, and what is the best strategy to deal with it?	What is happening now, and what have we done in the past to deal with it?
Use of descriptive analytics as a major component of analysis	Yes	Yes	Yes
Use of predictive analytics as a major component of analysis	Yes	Yes	No (only historically)
Use of prescriptive analytics as a major component of analysis	Yes	Yes	No (only historically)
Use of all three in combination	No	Yes	No
Business focus	Maybe	Yes	Yes
Focus of storing and maintaining data	No	No	Yes
Required focus of improving business value and performance	No	Yes	No

1.2 Business Analytics Process

The complete *business analytics process* involves the three major component steps applied sequentially to a source of data (see Figure 1.1). The outcome of the business analytics process must relate to business and seek to improve business performance in some way.

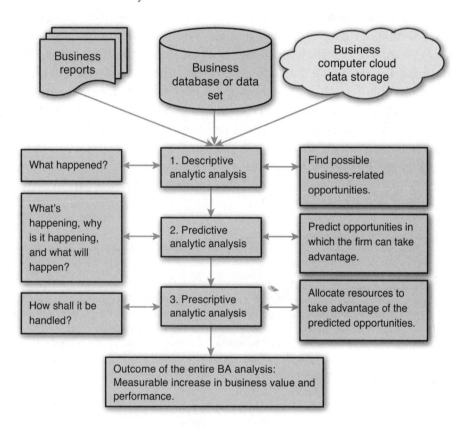

Figure 1.1 Business analytics process

The logic of the BA process in Figure 1.1 is initially based on a question: What valuable or problem-solving information is locked up in the sources of data that an organization has available? At each of the three steps that make up the BA process, additional questions need to be answered, as shown in Figure 1.1. Answering all these questions requires mining the information out of the data via the three steps of analysis that comprise the BA process. The analogy of digging in a mine is appropriate for the BA process because finding new, unique, and valuable information that can lead to a successful strategy is just as good as finding gold in a mine. SAS, a major analytic corporation (www.sas.com), actually has a step in its BA process, *Query Drill-down*, which refers to the mining effort of questioning and finding answers to pull up useful information in the BA analysis. Many firms routinely undertake BA to solve specific problems, whereas other firms undertake BA to explore and discover new knowledge to guide organizational planning and decision-making to improve business performance.

The size of some data sources can be unmanageable, overly complex, and generally confusing. Sorting out data and trying to make sense of its informational value requires the application of descriptive analytics as a first step in the BA process. One might begin simply by sorting the data into groups using the four possible classifications presented in Table 1.4. Also, incorporating some of the data into spreadsheets like Excel and preparing cross tabulations and contingency tables are means of restricting the data into a more manageable data structure. Simple measures of central tendency and dispersion might be computed to try to capture possible opportunities for business improvement. Other descriptive analytic summarization methods, including charting, plotting, and graphing, can help decision makers visualize the data to better understand content opportunities.

Table 1.4 Types of Data Measurement Classification Scales

Type of Data Measurement Scale	Description
Categorical Data	Data that is grouped by one or more characteristics. Categorical data usually involves cardinal numbers counted or expressed as percentages. Example 1: Product markets that can be characterized by categories of "high-end" products or "low-income" products, based on dollar sales. It is common to use this term to apply to data sets that contain items identified by categories as well as observations summarized in cross-tabulations or contingency tables.
Ordinal Data	Data that is ranked or ordered to show relational preference. Example 1: Football team rankings not based on points scored but on wins. Example 2: Ranking of business firms based on product quality.
Interval Data	Data that is arranged along a scale, in which each value is equally distant from others. It is ordinal data. Example 1: A temperature gauge. Example 2: A survey instrument using a Likert scale (that is, 1, 2, 3, 4, 5, 6, 7), where 1 to 2 is perceived as equidistant to the interval from 2 to 3, and so on. Note: In ordinal data, the ranking of firms might vary greatly from first place to second, but in interval data, they would have to be relationally proportional.
Ratio Data	Data expressed as a ratio on a continuous scale. Example 1: The ratio of firms with green manufacturing programs is twice that of firms without such a program.

From Step 1 in the *Descriptive Analytic analysis* (see Figure 1.1), some patterns or variables of business behavior should be identified representing targets of business opportunities and possible (but not yet defined) future trend behavior. Additional effort (more mining) might be required, such as the generation of detailed statistical reports narrowly focused on the data related to targets of business opportunities to explain what is taking place in the data (what happened in the past). This is like a statistical search for predictive variables in data that may lead to patterns of behavior a firm might take advantage of if the patterns of behavior occur in the future. For example, a firm might find in its general sales information that during economic downtimes, certain products are sold to customers of a particular income level if certain advertising is undertaken. The sales, customers, and advertising variables may be in the form of any of the measurable scales for data in Table 1.4, but they have to meet the three conditions of BA previously mentioned: clear relevancy to business, an implementable resulting insight, and performance and value measurement capabilities.

To determine whether observed trends and behavior found in the relationships of the descriptive analysis of Step 1 actually exist or hold true and can be used to forecast or predict the future, more advanced analysis is undertaken in Step 2, *Predictive Analytic analysis*, of the BA process. There are many methods that can be used in this step of the BA process. A commonly used methodology is multiple regression. (See Appendix A, "Statistical Tools," and Appendix E, "Forecasting," for a discussion on multiple regression and ANOVA testing.) This methodology is ideal for establishing whether a statistical relationship exists between the predictive variables found in the descriptive analysis. The relationship might be to show that a dependent variable is predictively associated with business value or performance of some kind. For example, a firm might want to determine which of several promotion efforts (independent variables measured and represented in the model by dollars in TV ads, radio ads, personal selling, or magazine ads) is most efficient in generating customer sales dollars (the dependent variable and a measure of business performance). Care would have to be taken to ensure the multiple regression model was used in a valid and reliable way, which is why ANOVA and other statistical confirmatory analyses support the model development. Exploring a database using advanced statistical procedures to verify and confirm the best predictive variables is an important part of this step in the BA process. This answers the questions of what is currently happening and why it happened between the variables in the model.

A single or multiple regression model can often forecast a trend line into the future. When regression is not practical, other forecasting methods (exponential smoothing, smoothing averages) can be applied as predictive analytics to develop needed forecasts of business trends. (See Appendix E.) The identification of future trends is the main output of Step 2 and the predictive analytics used to find them. This helps answer the question of what will happen.

If a firm knows where the future lies by forecasting trends as they would in Step 2 of the BA process, it can then take advantage of any possible opportunities predicted in that future state. In Step 3, *Prescriptive Analytics analysis*, operations research methodologies can be used to optimally allocate a firm's limited resources to take best advantage of the opportunities it found in the predicted future trends. Limits on human, technology, and financial resources prevent any firm from going after all opportunities it may have available at any one time. Using prescriptive analytics allows the firm to allocate limited resources to optimally achieve objectives as fully as possible. For example, *linear programming* (a constrained optimization methodology) has been used to maximize the profit in the design of supply chains (Paksoy et al., 2013). (Note: Linear programming and other optimization methods are presented in Appendixes B, "Linear Programming," C, "Duality and Sensitivity Analysis in Linear Programming," and D, "Integer Programming.") This third step in the BA process answers the question of how best to allocate and manage decision-making in the future.

In summary, the three major components of descriptive, predictive, and prescriptive analytics arranged as steps in the BA process can help a firm find opportunities in data, predict trends that forecast future opportunities, and aid in selecting a course of action that optimizes the firm's allocation of resources to maximize value and performance. The BA process, along with various methodologies, will be detailed in Chapters 5 through 10.

1.3 Relationship of BA Process and Organization Decision-Making Process

The BA process can solve problems and identify opportunities to improve business performance. In the process, organizations may also determine strategies to guide operations and help achieve competitive advantages. Typically, solving problems and identifying strategic opportunities to follow are organization decision-making tasks. The latter, identifying opportunities, can be viewed as a problem of strategy choice requiring a solution. It should come as no surprise that the BA process described in Section 1.2 closely parallels classic organization decision-making processes. As depicted in Figure 1.2, the business analytics process has an inherent relationship to the steps in typical organization decision-making processes.

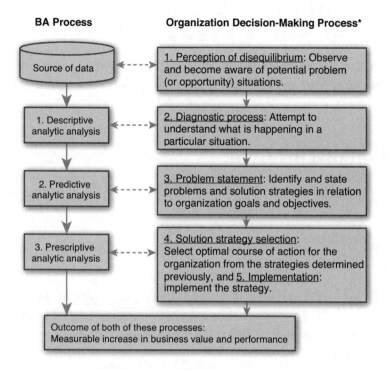

BA Process   **Organization Decision-Making Process***

Source of data ⟵----⟶ 1. Perception of disequilibrium: Observe and become aware of potential problem (or opportunity) situations.

1. Descriptive analytic analysis ⟵----⟶ 2. Diagnostic process: Attempt to understand what is happening in a particular situation.

2. Predictive analytic analysis ⟵----⟶ 3. Problem statement: Identify and state problems and solution strategies in relation to organization goals and objectives.

3. Prescriptive analytic analysis ⟵----⟶ 4. Solution strategy selection: Select optimal course of action for the organization from the strategies determined previously, and 5. Implementation: implement the strategy.

Outcome of both of these processes: Measurable increase in business value and performance

Figure 1.2 Comparison of business analytics and organization decision-making processes

Source: Adapted from Figure 1 in Elbing (1970), pp. 12–13.

The *organization decision-making process* (ODMP) developed by Elbing (1970) and presented in Figure 1.2 is focused on decision-making to solve problems but could also be applied to finding opportunities in data and deciding what is the best course of action to take advantage of them. The five-step ODMP begins with the perception of disequilibrium, or the awareness that a problem exists that needs a decision. Similarly, in the BA process, the first step is to recognize that databases may contain information that could both solve problems and find opportunities to improve business performance. Then in Step 2 of the ODMP, an exploration of the problem to determine its size, impact, and other factors is undertaken to diagnose what the problem is. Likewise, the BA descriptive analytic analysis explores factors that might prove useful in solving problems and offering opportunities. The ODMP problem statement step is similarly structured to the BA predictive analysis to find strategies, paths, or trends that clearly define a problem or opportunity for an organization to solve problems. Finally, the ODMP's last steps of strategy selection and implementation involve the same kinds of tasks that the BA process requires in the final prescriptive step (make an optimal selection of resource allocations that can be implemented for the betterment of the organization).

The decision-making foundation that has served ODMP for many decades parallels the BA process. The same logic serves both processes and supports organization decision-making skills and capacities.

1.4 Organization of This Book

This book is designed to answer three questions about BA:

- What is it?
- Why is it important?
- How do you do it?

To answer these three questions, the book is divided into three parts. In Part I, "What Is Business Analytics?" Chapter 1 answers the "what" question. In Part II, the "why" question is answered in Chapter 2, "Why Is Business Analytics Important?" and Chapter 3, "What Resource Considerations Are Important to Support Business Analytics?"

Knowing the importance of explaining how BA is undertaken, the rest of the book's chapters and appendixes are devoted to answering that question. Chapter 4, "How Do We Align Resources to Support Business Analytics within an Organization?" explains how an organization needs to support BA. Chapter 5, "What Is Descriptive Analytics?" Chapter 6, "What Is Predictive Analytics?" and Chapter 7, "What Is Prescriptive Analytics?" detail and illustrate the three respective steps in the BA process. To further illustrate how to conduct a BA analysis, Chapter 8, "A Final Business Analytics Case Problem," provides an example of BA. Supporting the analytic discussions is a series of analytically oriented appendixes that follow Chapter 8.

Part III, "How Can Business Analytics Be Applied?" includes quantitative analyses utilizing computer software. In an effort to provide some diversity of software usage, SAS and LINGO software output are presented. Because of the changing nature of software and differing educational backgrounds, this book does not provide extensive software explanation.

In addition to the basic content that makes up the body of the chapters, there are pedagogy enhancements that can aid learning. All chapters begin with chapter objectives and end with a summary, discussion questions, and, where needed, references. In addition, Chapters 5 through 8 have sample problems with solutions, as well as additional assignment problems.

Some of the more detailed explanations of methodologies are presented in the appendixes. Their positioning in the appendixes is designed to enhance content flow and permit more experienced readers a flexible way to select only the technical content they might want to use. An extensive index allows quick access to terminology.

Summary

This chapter has introduced important terminology and defined business analytics in terms of a unique process useful in securing information on which decisions can be made and business opportunities seized. Data classification measurement scales were also briefly introduced to aid in understanding the types of measures that can be employed in BA. The relationship of the BA process and the organization decision-making process was explained in terms of how they complement each other. This chapter ended with a brief overview of this book's organization and how it is structured to aid learning.

Knowing *what* business analytics is about is important, but equally important is knowing *why* it is important. Chapter 2 begins to answer the question.

Discussion Questions

1. What is the difference between analytics and business analytics?
2. What is the difference between business analytics and business intelligence?
3. Why are the steps in the business analytics process sequential?
4. How is the business analytics process similar to the organization decision-making process?
5. Why does interval data have to be relationally proportional?

References

Bartlett, R. (2013). *A Practitioner's Guide to Business Analytics*. McGraw-Hill, New York, NY.

Elbing, A. O. (1970). *Behavioral Decisions in Organizations*. Scott Foresman and Company, Glenview, IL.

Isson, J. P., Harriott, J. S. (2013). *Win with Advanced Business Analytics*. John Wiley & Sons, Hoboken, NJ.

Negash, S. (2004). "Business Intelligence." *Communications of the Association of Information Systems*. Vol. 13, pp. 177–195.

Paksoy, T., Ozxeylan, E., Weber, G. W. (2013). "Profit-Oriented Supply Chain Network Optimization." *Central European Journal of Operations Research*. Vol. 21, No. 2, pp. 455–478.

Stubbs, E. (2011). *The Value of Business Analytics*. John Wiley & Sons, Hoboken, NJ.

Part II

Why Is Business Analytics Important?

2

Why Is Business Analytics Important?

Chapter objectives:

- Explain why business analytics is important in solving business problems.
- Explain why business analytics is important in identifying new business initiatives.
- Describe the kinds of questions business analytics can help answer.
- Explain how business analytics can help an organization achieve a competitive advantage.
- Explain different types of competitive advantages and their relationship to business analytics.
- Explain the importance of business analytics for a business organization.

2.1 Introduction

Telecommunication and information systems are collecting data on every aspect of life with incredible rates of speed and comprehensiveness. In addition, businesses are running opinion surveys and collecting all forms of data for their operations. With information system clouds providing large amounts of data that are easily available and data warehousing systems capable of storing big data in large databases, there is presently a need to process information out of data to gain knowledge and justify data investment. As Demirkan and Delen (2013) have shown, placing large data into computer clouds can provide business analytics in a timely and agile way. Firms recognize the need for this information to be competitive, and business analytics is one strategy to gain the knowledge they seek.

The problem with big data or even small data files is that they can easily obscure the information desired. Sometimes a small alteration in a piece of data located in a file can change meanings. The 1960s television program *The Prisoner* used the catchphrase, "I want information." When this phrase is seen in print or spoken, it denotes that someone wants information. Yet when the term was used in *The Prisoner*, it referred to "in" and "formation." (That is, "I want in formation.") The phrase was used to make the prisoner do what he was told and act like the others. Note that a single space in this second phrase completely changes the meaning. Mining for relevant business information in big databases when small differences can alter meanings makes it a challenge to find relevant and useful information. Business analytics as a process is designed to meet this challenge.

2.2 Why BA Is Important: Providing Answers to Questions

It may seem overly virtuous, but BA is the next best thing to a crystal ball for answering important business questions. In each of the three steps of the BA process (from Chapter 1, "What Is Business Analytics?"), answers to a variety of questions can and should be answered as a logical outcome of the analysis. The answers become the basis of information and knowledge that makes BA a valued tool for decision-making and helps explain why it is important to learn and use.

As can be seen in Table 2.1, the sampling of the kinds of questions a typical BA analysis can render is related not only to each step in the BA process, but to the context of time. To better understand the value of the information BA analysis provides and understand why this subject is important to improved business performance, a simple illustrative case scenario is presented.

Table 2.1 Questions Business Analytics Seeks to Answer*

	Time Period		
Step in BA	**Past**	**Present**	**Future**
1. Descriptive	What happened in the past?	What is happening now based on the past?	What will appear to happen based on the past?
2. Predictive	How did it happen in the past? Why did it happen in the past?	What possible trends exist in the data that can predict or forecast what course of action should be taken now?	What is the range and likelihood of possible outcomes that can happen if the current trends or forecasts are allowed?
3. Prescriptive	How best can we leverage what we know from the trends and forecasts?	How can we optimally apply resources to maximize the business performance outcomes in the future?	How can we continuously apply BA in the future to optimize upcoming business performance outcomes?

*Source: Adapted from Exhibit 9.1 from Isson and Harriott (2013), p. 169.

In this illustrative case scenario, a local credit union offers a series of packaged homeowner loans that are periodically marketed by running a promotional campaign in a variety of media (print ads, television commercials, radio spots). The idea is to bring in new customers to make home loans that fit one of the packaged deals. Halfway through the marketing program, the credit union does not know if the business generated is due to the promotional campaign or just a result of its normal business cycle. To clarify, the credit union undertakes a BA analysis. The resulting information from the BA analysis (based on the same questions as in Table 2.1) is presented in Table 2.2. Reading first the Descriptive step, Past, Present, and Future, and then sequentially following the same pattern with the Predictive and Prescriptive steps, the possible types of information gleaned from these BA questions and answers can be illustrated by this example.

Table 2.2 Credit Union Example of BA Analysis Information

	Time Period		
Step in BA	**Past**	**Present**	**Future**
1. Descriptive	Based on graphics results, past ad campaigns resulted in a moderate increase in new loans.	Based on sorting of loan activities, new homeowner loans are experiencing just a moderate increase in new loan applications.	Based on histogram of loans to date, there is no discernible pattern, just uniform new loan sales that are constant over time. No business cycle impact is observed to alter loan patterns.
2. Predictive	Statistically, correlations have revealed in the past that marketing promotions will increase new loans, but why they generate new loans depends on how the promotion campaign invests its funding.	Utilizing multiple regression, the model predicts that a greater allocation in funds for television commercials and print ads will be more effective in generating new loans than investing in radio spots.	Utilizing variance statistics from the regression model, a confidence interval can estimate the number of new loans possible if a reallocation of promotion funds is implemented.
3. Prescriptive	Reallocating marketing budget funds from radio spots to television commercials and print ads is required to more effectively reach the target audience.	Given the constrained resource of funding, a linear programming model is used to optimally allocate the marketing budget in dollars to maximize the promotional outcome for new loans.	Track new loan applications caused by the promotion campaign. Map actual results to the predicted outcome suggested in the analysis.

The answers to the questions raised in the credit union example are typical of any business organization problem-solving or opportunity-seeking quest. The answers were not obtained by just using statistics, computer search routines, or operations research methodologies, but rather were a result of a sequential BA process. The informational value of the answers in this scenario suggests a measurable and precise course of action for the management of the credit union to follow. By continuously applying BA as a decision support system, firms have come to see not only why they need BA, but also how BA can become a strategy to achieve competitive advantage. Kiron et al. (2012) reported in a survey on business through the year 2012 that firms applying business analytics permit the organization to have better access to data for decision-making and offer a competitive advantage.

2.3 Why BA Is Important: Strategy for Competitive Advantage

Companies that make plans that generate successful outcomes are winners in the marketplace. Companies that do not effectively plan tend to be losers in the marketplace. Planning is a critical part of running any business. If it is done right, it obtains the results that the planners desire.

Business organization planning is typically segmented into three types, presented in Figure 2.1. The planning process usually follows a sequence from strategy, down to tactical, and then down to operational, although Figure 2.1 shows arrows of activities going up and down the depicted hierarchical structure of most business organizations. The upward flow in Figure 2.1 represents the information passed from lower levels up, and the downward flow represents the orders that are passed from higher levels of management down to lower levels for implementation. It can be seen in the Teece (2007) study and more recently in Rha (2013) that the three steps in the BA process and strategic planning embody the same efforts and steps.

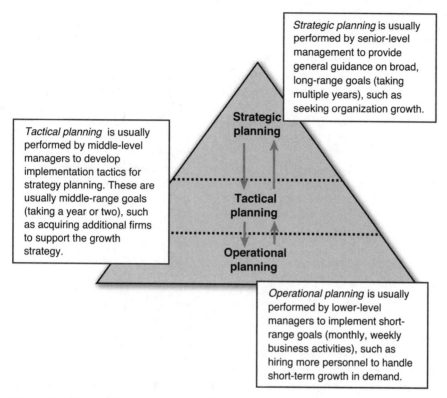

Figure 2.1 Types of organization planning*

*Source: Adapted from Figure 1.2 in Schniederjans and LeGrand (2013), p.9.

Effectively planning and passing down the right orders in hopes of being a business winner requires good information on which orders can be decided. Some information can become so valuable that it provides the firm a *competitive advantage* (the ability of one business to perform at a higher level, staying ahead of present competitors in the same industry or market). Business analytics can support all three types of planning with useful information that can give a firm a competitive advantage. Examples of the ways BA can help firms achieve a competitive advantage are presented in Table 2.3.

Table 2.3 Ways BA Can Help Achieve a Competitive Advantage

Type of Competitive Advantage	Description	Ways BA Can Help Achieve the Competitive Advantage
Price Leadership	From a marketing standpoint, offer products or services at the lowest cost to customers in the industry, while making acceptable profit for the company.	Identify main competitors; monitor, report, and accurately forecast competitive prices so firm can keep lowest cost profile while maintaining and measuring profit margins.
Sustainability	To ensure the firm's resource usage in a way that seeks balance to hurt neither the environment nor the bottom line of a firm's profitability.	Identify areas needing resource reallocations to avoid damaging the environment, suggest ways to reallocate the resources, and help allocate them optimally to achieve the best possible balance.
Operations Efficiency	Improve the internal business operations and activities over competitors, lessening the cost to the customer. That reduced cost, if passed on to customers, can provide a lower price advantage based on efficiency.	Identify operation areas needing correction or modification and suggest alternatives to improve efficiency. Also, this can be useful in selecting which alternative to use to maximize business performance.
Service Effectiveness	Make customer transactions easier or more pleasurable than with other firms. This improves the service characteristics of the firm while decreasing the time it takes to get services to the customer, thus enhancing customer value.	Obtain customer opinions on problem service areas needing fixing; explain why the fix is needed, suggest alternatives to the fix, improve the effectiveness of the service operations, and measure and report improvements.
Innovation	Introduce completely new or notably better products or services with the intention of disrupting competitors' businesses by obsoleting the current market entries with breakthrough product offerings.	Obtain and validate customer ideas and suggestions on new products or enhancements in current products. Monitor customer reactions and suggest refinements as new products are introduced to customers. Monitor performance on new products and report results.
Product Differentiation	Provide customers with a variety of products, services, or features that competitors are not yet offering or are unable to offer.	Identify new products not offered by competitors, suggest new services to offer, and forecast potential of new products for profitability measurement.

2.4 Other Reasons Why BA Is Important

There is an almost endless list of potential applications of BA to provide information on which decisions can be made or improved.

2.4.1 Applied Reasons Why BA Is Important

Some potential applications for decision-making will be illustrated in later chapters. Several brief examples are described in Table 2.4.

Table 2.4 Applications of BA to Enhance Decision-Making

Decision-Making Application	Description
Increasing Customer Profitability	BA can provide detailed information (current pricing and pricing trends) on competitor products. This information can be used to set prices that allow a firm to keep profit margins at a profit-maximizing level by balancing sales volume with lower prices and margins or increasing prices to increase margins, depending on competitor pricing.
Risk Reduction	With the types of information provided in all steps of the BA process, businesses do not have to guess but instead can be guided with some probabilistically computed likelihood of certainty on sales, budgets, and human and technology decisions. Having a probability estimate as a guide reduces the risk of poor judgments.
Merchandize Strategy Optimization	Quantitative tools used in the Prescriptive step of the BA process can be used to determine optimal layout designs for store merchandize, cost-minimizing inventory levels, and even scheduling for sales staff to help achieve maximum merchandising results.
Human Resource Decisions	According to Fitx-enz (2013, pp. 223–245) analytics determined on *human resources* (HR) in the Predictive step of the BA process can be computed (workforce productivity, revenue per *full-time equivalent* [FTE], cost per FTE, and so on). In turn, this can answer questions like what should be done to improve the quality of HR hires, what types of training would be most effective, and how incentive pay can be used to stimulate performance.

Decision-Making Application	Description
Business Performance Tracking	In addition to the normal data collection in the Descriptive step of the BA process, specific business performance parameters can be continually collected, monitored, and measured. Then analytics can update those measures to provide an up-to-date performance achievement index useful for comparing performance over time. In addition, the Predictive step of the BA forecasts of expected performance can be used to set planning and performance goals to guide operations.

2.4.2 The Importance of BA with New Sources of Data

As advances in new computer and telecommunication technologies take place, they provide new types of data. Therefore, new types of analytics need to be applied in BA analyses. *Digital analytics* is a term that describes any source of data that is conveyed using digital sources. Examples of these new sources of data-based analytics include text analytics and unstructured data analytics. *Text analytics* can be defined as a set of linguistic, statistical, and computer-based techniques that model and structure the information content from textual sources (Basole et al., 2013). It is a search process in databases to find patterned text material that provides useful information. Also referred to as *text data mining*, text analytics uses data mining software to look into databases to find and validate the kinds of information on which predictions can be made.

Being able to search and quantify textual data using text analytics opens great opportunities to glean information about customers and markets based on technology-driven data collection technologies. One example of technology-driven data is social media data. *Social media* can be defined as interactions or communications among people or communities, usually performed on a technology platform, involving the sharing, creating, discussing, and modifying of communicated verbal or electronic content. Two global social platforms are *Twitter* and *Facebook*. The methodologies or technologies used in the purveyance of social media data can include any means of distribution of verbal or other types of communications, including, but not limited to, photographs or pictures, video, Internet forums, web logs, discussion forums, social blogs, wikis, social networks, and podcasts. These sources of data are the basis of *social media analytics*, on which the analytics information can aid in learning new types of social media behavior and information. They provide a great challenge for BA analysts because of the excessive volume and difficulty in quantifying the information in useful ways. They also provide a great opportunity to find information that might create a competitive advantage. An example of how social media analytics helped find auto defects was illustrated in a study by Abrahams et al. (2012). By employing text mining on a social medium (online discussion forums) used by vehicle enthusiasts, a variety of quality defects were identified, categorized, and prioritized for automobile manufacturers to correct.

Another similar digital source of analytics is referred to as mobile analytics. *Mobile analytics* can be defined as any data secured from mobile devices, such as smartphones, iPhones, iPads, and Web browsers. All these are mobile technologies used to obtain digital data from the interaction of people (Evans, 2010). The fact that they are mobile and move from location to location with the user differentiates the type of information available to the analytics analysts. For example, the mobile technology allows analysts to not only track what a potential customer might talk about on the use of a product (such as in social media analytics), but track movements of where the customer makes decisions on products. That can help explain why those decisions are made. For example, mobile technology might reveal the location of a purchaser of hair spray to have been physically located near an area where billboards are used for hair spray advertising, thus helping to reveal the possible connection and effectiveness of a billboard promotion.

When data is placed in databases and can be logically filed, accessed, referenced, and used, it is known as *structured data*. When data or information, either digital or nondigital, cannot be put into a database or has no predefined structure, it is known as *unstructured data*. Examples of unstructured data include images, text, and other data that, for one reason or another, cannot be placed in a logically searchable database based on content. This data can be digitally stored in a file, but not in a way that can be usefully retrieved using any kind of logic model or sorting process. Much of the data contained in e-mails and on the Web is unstructured. Another way of looking at unstructured data is that it is what is left over and cannot be placed in a structured database. As time goes on, more effort in developing complex algorithms and other computer-based technologies will be applied to unstructured data, reducing the amount of data that falls into this category. Given the volume of graphics data or other unstructured data generated every day, the challenge to BA analysts will be an ever-growing effort to understand and structure the unstructured data that remains in an effort to gain its informational value. Part of the value and importance of BA is in accepting this challenge.

Summary

This chapter sought to explain why business analytics is an important subject for business organizations. It discussed how BA can answer important questions and how it can help a firm achieve a competitive advantage. In addition, it presented the role of BA in organization planning. Finally, it introduced other types of digital analytics to explain their beneficial role and challenges to BA.

We move in the next chapter to further explain why BA is important in the context of its required investment. Like any management task, the successful use of BA requires an investment in human and technology resources. Chapter 3, "What Resource Considerations Are Important to Support Business Analytics?" explores the allocation of resources to maximize BA performance and explains why the investment is needed.

Discussion Questions

1. Why does each step in the business analytics process have a past, present, and future dimension?
2. What is a competitive advantage, and how is it related to BA?
3. Why does having the ability to aid in decision-making make BA important?
4. How does BA help achieve sustainability?
5. What are digital analytics?

References

Abrahams, A., Jiao, J., Wang, G., Fan, W. (2012). "Vehicle Defect Discovery from Social Media." *Decision Support Systems*. Vol. 54, No. 1, pp. 87–97.

Basole, R., Seuss, C., Rouse, W. (2013). "IT Innovation Adoption by Enterprises: Knowledge Discovery through Text Analytics." *Decision Support Systems*. Vol. 54, No. 2, pp. 1044–1054.

Demirkan, H., Delen, D. (2013). "Leveraging the Capabilities of Service-Oriented Decision Support Systems: Putting Analytics and Big Data in Cloud." *Decision Support Systems*. Vol. 55, No. 1, pp. 412–421.

Evans, B. (2010). "The Rise of Analytics and Fall of the Tactical CIO." *Informationweek*. December 6, No. 1286, p. 14.

Fitx-enz, J. (2013). "Predictive Analytics Applied to Human Resources." In Isson, J. P., Harriott, J. S. (2013) *Win with Advanced Business Analytics*. John Wiley & Sons, Hoboken, NJ.

Isson, J. P., Harriott, J. S. (2013). *Win with Advanced Business Analytics*. John Wiley & Sons, Hoboken, NJ.

Kiron, D., Kirk-Prentice, P., Boucher-Ferguson, R. (2012). "Innovating with Analytics." *MIT Sloan Management Review*. Vol. 54, No. 1, pp. 47–52.

Rha, J. S. (2013). "Ambidextrous Supply Chain Management as a Dynamic Capability: Building a Resilient Supply Chain" (Doctoral Dissertation).

Schniederjans, M. J., LeGrand, S. B. (2013). *Reinventing the Supply Chain Life Cycle*. FT Press, Upper Saddle River, NJ.

Teece, D. J. (2007). "Explicating Dynamic Capabilities: The Nature and Microfoundations of (Sustainable) Enterprise Performance." *Strategic Management Journal*. Vol. 28, No. 13, pp. 1319–1350.

3

What Resource Considerations Are Important to Support Business Analytics?

Chapter objectives:

- Explain why personnel, data, and technology are needed in starting up a business analytics program.
- Explain what skills business analytics personnel should possess and why.
- Describe the job specialties that exist in business analytics.
- Describe database encyclopedia content.
- Explain the categorization of data in terms of sources.
- Describe internal and external sources of data.
- Describe an information technology infrastructure.
- Describe a database management system and how it supports business analytics.

3.1 Introduction

To fully understand why business analytics (BA) is necessary, one must understand the nature of the roles BA personnel perform. In addition, it is necessary to understand resource needs of a BA program to better comprehend the value of the information that BA provides. The need for BA resources varies by firm to meet particular decision support requirements. Some firms may choose to have a modest investment, whereas other firms may have BA teams or a department of BA specialists. Regardless of the level of resource investment, at minimum, a BA program requires resource investments in BA personnel, data, and technology.

3.2 Business Analytics Personnel

One way to identify personnel needed for BA staff is to examine what is required for certification in BA by organizations that provide BA services. *INFORMS* (www. informs.org/Certification-Continuing-Ed/Analytics-Certification), a major academic and professional organization, announced the startup of a *Certified Analytic Professional* (CAP) program in 2013. Another more established organization, *Cognizure* (www.cognizure.com/index.aspx), offers a variety of service products, including business analytics services. It offers a general certification *Business Analytics Professional* (BAP) exam that measures existing skill sets in BA staff and identifies areas needing improvement (www.cognizure.com/cert/bap.aspx). This is a tool to validate technical proficiency, expertise, and professional standards in BA. The certification consists of three exams covering the content areas listed in Table 3.1.

Table 3.1 Cognizure Organization Certification Exam Content Areas*

Exam	Topic	Specific Content Areas Covered	Examples
I	Statistical Methods	1. Visualizing and Exploring Data 2. Descriptive Statistics 3. Probability Distributions 4. Sampling and Estimation 5. Statistical Inference 6. Regression Analysis 7. Predictive Modeling and Analysis	1. Graphs and charts 2. Mean, median, mode 3. Normal distribution 4. Confidence intervals 5. Hypothesis testing 6. Multiple regression 7. Curve fitting of models and functions to raw data
II	Operations Research Methods	1. Linear Optimization 2. Integer Optimization 3. Nonlinear Optimization 4. Simulation 5. Decision Analysis 6. Forecasting	1. Linear programming 2. Integer programming 3. Quadratic programming 4. Monte Carlo method 5. Expected value analysis 6. Exponential smoothing
III	Case Studies	Practical knowledge of real-world situations	Application of the preceding methods to solve a real-world problem

Source: Adapted from Cognizure Organization website (www.cognizure.com/cert/bap.aspx).

Most of the content areas in Table 3.1 will be discussed and illustrated in subsequent chapters and appendixes. The three exams required in the Cognizure certification program can easily be understood in the context of the three steps of the BA process (descriptive, predictive, and prescriptive) discussed in previous chapters. The topics in Figure 3.1 of the certification program are applicable to the three major steps in the BA process. The basic statistical tools apply to the descriptive analytics step, the more advanced statistical tools apply to the predictive analytics step, and the operations research tools apply to the prescriptive analytics step. Some of the tools can be applied to both the descriptive and the predictive steps. Likewise, tools like simulation can be applied to answer questions in both the predictive and the prescriptive steps, depending on how they're used. At the conjunction of all the tools is the reality of case studies. The use of case studies is designed to provide practical experience, whereby all tools are employed to answer important questions or seek opportunities.

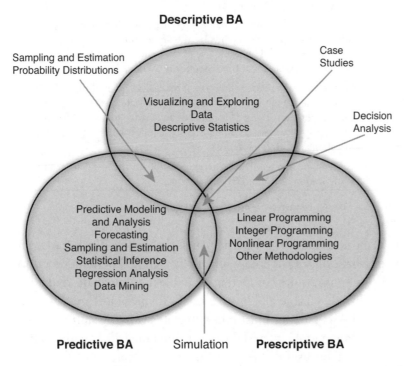

Figure 3.1 Certification content areas and their relationship to the steps in BA

Other organizations also offer specialized certification programs. These certifications include other areas of knowledge and skills beyond just analytic tools. IBM, for example, offers a variety of specialized BA certifications (www-03.ibm.com/certify/certs/ba_index.shtml). Although these include certifications in several dozen statistical, information systems, and analytic methodologies related to BA, they also include specialized skill sets related to BA personnel (administrators, designers, developers, solution experts, and specialists), as presented in Table 3.2.

Table 3.2 Types of BA Personnel*

BA Personnel Specialty	Description
Administrators	Within the context of the IBM BA and business intelligence (BI) software platforms, administrators manage servers (their load balancing, installation, and configurations). They manage reports from computer portals, manage dispatchers, and perform troubleshooting for technology. They are also in charge of user authorization and authentication for security.
Designers	As members of a team, designers are responsible for building reports using relational data models, as well as enhancing, customizing, and managing professional reports.
Developers	As members of a team, developers require skills that are closely tied to the BA process and involve the application of analytics, data warehousing, model building, use of operations research and statistical methodologies, and real-time monitoring of data flow to users.
Solution Experts	As members of a team, solution experts analyze, plan, design, deploy, and operate BA applications using an appropriate methodology and development approach. This requires knowledge in many differing BA software applications, including statistical, information systems, and operations research methods.
Technical Specialists	As members of a team, they are responsible for the installation and configuration of BA and BI applications.

*Source: Adapted from IBM website (www-03.ibm.com/certify/certs/ba_index.shtml).

With the variety of positions and roles participants play in the BA process, this leads to the question of what skill sets or competencies are needed to function in BA. In a general sense, BA positions require competencies in business, analytic, and information systems skills. As listed in Table 3.3, business skills involve basic management of people and processes. BA personnel must communicate with BA staffers within the organization (the BA team members) and the other functional areas within a firm (BA customers and users) to be useful. Because they serve a variety of functional areas within a firm, BA personnel need to possess customer service skills so they can interact with the firm's personnel and understand the nature of the problems they seek to solve. BA personnel also need to sell their services to users inside the firm. In addition, some must lead a BA team or department, which requires considerable interpersonal management leadership skills and abilities.

Table 3.3 Select Types of BA Personnel Skills or Competency Requirements

Type of Skill or Competency	Description of Possible Roles
Business	• Leadership • People-related management and communication skills • Manage BA projects (prioritize, schedule, and so on) • Manage BA processes (rules, procedures governing acceptance and use) • Determine project requirements • Train BA personnel to develop competencies
Analytic	• Know how to use statistical methodologies • Know how to use operations research methodologies • Know how to use data mining for quantitative data and text analytics for unstructured data
Information system	• Maintain and use computer portals • Identify and extract data • Maintain quality data

Fundamental to BA is an understanding of analytic methodologies listed in Table 3.1 and others not listed. In addition to any tool sets, there is a need to know how they are integrated into the BA process to leverage data (structured or unstructured) and obtain information that customers who will be guided by the analytics desire.

In summary, people who undertake a career in BA are expected to know how to interact with people and utilize the necessary analytic tools to leverage data into useful information that can be processed, stored, and shared in information systems in a way that guides a firm to higher levels of business performance.

3.3 Business Analytics Data

Structured and unstructured data (introduced in Chapter 2, "Why Is Business Analytics Important?") is needed to generate analytics. As a beginning for organizing data into an understandable framework, statisticians usually categorize data into meaning groups.

3.3.1 Categorizing Data

There are many ways to categorize business analytics data. Data is commonly categorized by either internal or external sources (Bartlett, 2013, pp. 238–239). Typical examples of internal data sources include those presented in Table 3.4. When firms try to solve internal production or service operations problems, internally sourced data may be all that is needed. Typical external sources of data (see Table 3.5) are numerous and provide great diversity and unique challenges for BA to process. Data can be measured quantitatively (for example, sales dollars) or qualitatively by *preference surveys* (for example, products compared based on consumers preferring one product over another) or by the amount of consumer discussion (chatter) on the Web regarding the pluses and minuses of competing products.

Table 3.4 Typical Internal Sources of Data on Which Business Analytics Can Be Based

Type of Internal Data	Description
Billing and Reminder Systems	Billing systems and reminder systems print bills and monitor customer payment information on valued-based customer segments.
Business	Industry codes, accounting information, personnel information, and so on are routinely collected in the course of business.
Customer	Names, addresses, returns, special contracts, segmentations, and so on are obtained when customers sign for or pay for products or services.
Customer Relationship Management Systems	*Customer relationship management* (CRM) systems collect and provide data on customer history, behavior on matters like complaints, the end of a relationship with a firm, and so on.
Human Resources	Information about employees, salaries, competencies, and so on is recorded by routine efforts over the history of employment.
Information from Enterprise Resource Planning Systems	*Enterprise resource planning* (ERP) systems are used to communicate internal business transactions to provide a direct feed of information on management issues and concerns, as well as other operations activities required to produce and sell products.
Product	Information is collected from procurement through post sales to monitor profitability, durability, and quality.
Production	Information that can be used to optimize production, inventory control, and supply chain delivery of the product to the customers is collected during the production processes.
Questionnaires	Information on customer behavior is obtained by customer questionnaires to measure customer service and product quality, among other things.
Web Logs	Information is collected on the firm's website usage via cookies and other means to learn customer navigation behavior and product interests.

Table 3.5 Typical External Sources of Data on Which Business Analytics Can Be Based

Type of External Data	Measured By
Customer Satisfaction	• Revenue, profit • Market share, sales • Product/service survey data • Loyalty • Brand awareness • Average spend per customer
Customer Demographics	• Geographic location (distance from market) • Income level • Market size
Competition	• Market share • Competitor profitability • Advertising/promotion efforts • Preference surveys • Web chatter on products
Economic	• Population statistics • Income distribution statistics

A major portion of the external data sources are found in the literature. For example, the *US Census* and the *International Monetary Fund* (IMF) are useful data sources at the macroeconomic level for model building. Likewise, audience and survey data sources might include *Nielsen* (www.nielsen.com/us/en.html) for psychographic or demographic data, financial data from *Equifax* (www.equifax.com), Dun & Bradstreet (www.dnb.com), and so forth.

3.3.2 Data Issues

Regardless of the source of data, it has to be put into a structure that makes it usable by BA personnel. We will discuss data warehousing in the next section, but here we focus on a couple of data issues that are critical to the usability of any database or data file. Those issues are data quality and data privacy. *Data quality* can be defined as data that serves the purpose for which it is collected. It means different things for different applications, but there are some commonalities of high-quality data. These qualities usually include accurately representing reality, measuring what it is supposed to measure, being timeless, and having completeness. When data is of high quality, it helps ensure competitiveness, aids customer service, and improves profitability. When data is of poor quality, it can provide information that is contradictory, leading to misguided decision-making. For example, having missing data in files can prohibit some forms' statistical modeling, and incorrect coding of information can completely render databases useless. Data quality requires effort on the part of data managers to cleanse data of erroneous information and repair or replace missing data. We will discuss some of these quality data measures in later chapters.

Data privacy refers to the protection of shared data such that access is permitted only to those users for whom it is intended. It is a security issue that requires balancing the need to know with the risks of sharing too much. There are many risks in leaving unrestricted access to a company's database. For example, competitors can steal a firm's customers by accessing addresses. Data leaks on product quality failures can damage brand image, and customers can become distrustful of a firm that shares information given in confidence. To avoid these issues, a firm needs to abide by the current legislation regarding customer privacy and develop a program devoted to data privacy.

Collecting and retrieving data and computing analytics requires the use of computers and information technology. A large part of what BA personnel do is related to managing information systems to collect, process, store, and retrieve data from various sources.

3.4 Business Analytics Technology

Firms need an *information technology* (IT) *infrastructure* that supports personnel in the conduct of their daily business operations. The general requirements for such a system are stated in Table 3.6. These types of technology are elemental needs for business analytics operations.

Table 3.6 General Information Technology (IT) Infrastructure

Type of Technology	Description
Computer Hardware	This is physical equipment used for input, processing, and output activities in an information system. Hardware can include computers of various sizes; various input, output, and storage devices; and telecommunications devices that link computers, including mobile handheld devices.
Computer Software	These are the preprogrammed instructions that control and coordinate the computer hardware components in the information system. They include system-wide software like ERP and smaller *apps* (computer software applications) for mobile devices.

Type of Technology	Description
Networking and Telecommunications Technology	Physical devices and software link the various pieces of hardware and transfer data from one physical location to another. They include the computers and communications equipment connected in networks for sharing voice, data, images, sound, and video. They also include the Internet, *intranets* (internal corporate networks based on Internet technology with limited access to employees within the firm), and *extranets* (private intranets extended to authorized users outside the organization).
Data Management Technology	Software governs the organization of data on physical storage media. It includes database management systems, data warehouses, data marts, and online analytical processing, as well as data, text, and Web mining technologies.

Of particular importance for BA is the data management technologies listed in Table 3.6. *Database management system* (DBMS) is a data management technology software that permits firms to centralize data, manage it efficiently, and provide access to stored data by application programs. DBMS usually serves as an interface between application programs and the physical data files of structured data. DBMS makes the task of understanding where and how the data is actually stored more efficient. In addition, other DBMS systems can handle unstructured data. For example, *object-oriented DBMS systems* are able to store and retrieve unstructured data, like drawings, images, photographs, and voice data. These types of technology are necessary to handle the load of big data that most firms currently collect.

DBMS includes capabilities and tools for organizing, managing, and accessing data in databases. Four of the more important capabilities are its data definition language, data dictionary, database encyclopedia, and data manipulation language. DBMS has a *data definition* capability to specify the structure of content in a database. This is used to create database tables and characteristics used in fields to identify content. These tables and characteristics are critical success factors for search efforts as the database grows. These characteristics are documented in the *data dictionary* (an automated or manual file that stores the size, descriptions, format, and other properties needed to characterize data). The *database encyclopedia* is a table of contents listing a firm's current data inventory and the data files that can be built or purchased. The typical content of the database encyclopedia is presented in Table 3.7. Of particular importance for BA are the *data manipulation language* tools included in DMBS. These tools are used to search databases for specific information. An example is *structure query language* (SQL), which allows users to find specific data through a session of queries and responses in a database.

Table 3.7 Database Encyclopedia Content

Database Content Item	Description
Purpose	Why the database exists, including any additional reports or analyses used in leveraging the data.
Time	Window of time when the data is collected or will be useful.
Source	Internal (auditing, accounting, and so on) and external (customers, and so on) sources.
Schematics	Diagrams illustrating the connections between tables and other data files.
Cost	Expense of collecting data, including purchasing prices.
Availability of Data	Window of time when the data may be available.
Collection Techniques	Methods of collection, including observation, data mining, census, and focus groups.
Collection Tools	Web, customer generated, e-survey, and so on.

Data warehouses are databases that store current and historical data of potential interest to decision makers. What a data warehouse does is make data available to anyone who needs access to it. In a data warehouse, the data is prohibited from being altered. Data warehouses also provide a set of query tools, analytical tools, and graphical reporting facilities. Some firms use intranet portals to make data warehouse information widely available throughout a firm.

Data marts are focused subsets or smaller groupings within a data warehouse. Firms often build enterprise-wide data warehouses in which a central data warehouse serves the entire organization and smaller, decentralized data warehouses (called data marts) are focused on a limited portion of the organization's data that is placed in a separate database for a specific population of users. For example, a firm might develop a smaller database on just product quality to focus efforts on quality customer and product issues. A data mart can be constructed more quickly and at lower cost than enterprise-wide data warehouses to concentrate effort in areas of greatest concern.

Once data has been captured and placed into database management systems, it is available for analysis with BA tools, including online analytical processing, as well as data, text, and Web mining technologies. *Online analytical processing* (OLAP) is software that allows users to view data in multiple dimensions. For example, employees can be viewed in terms of their age, sex, geographic location, and so on. OLAP would allow identification of the number of employees who are age 35, male, and in the western region of a country. OLAP allows users to obtain online answers to ad hoc questions quickly, even when the data is stored in very large databases.

Data mining is the application of a software discovery-driven process that provides insights into business data by finding hidden patterns and relationships in big data or large databases and inferring rules from them to predict future behavior. The observed patterns and rules guide decision-making. They can also act to forecast the impact of those decisions. It is an ideal predictive analytics tool used in the BA process mentioned in Chapter 1, "What Is Business Analytics?" The kinds of information obtained by data mining include those in Table 3.8.

Table 3.8 Types of Information Obtainable with Data Mining Technology

Types of Information	Description	Example
Associations	Occurrences linked to a single event.	An ad in a newspaper is associated with greater sales.
Classification	Recognizes patterns that describe the group an item belongs to by examining previous classified existing items and by inferring a set of rules that guide the classification process.	Identify customers who are likely to need more customer service than those who need less.
Clustering	Similar to classification when no groups have yet been defined, helps to discover different groupings within data.	Identify groups that can be differentiated within a single, large group of customers. An example would be distinguishing tea drinkers from those who would drink other beverages offered in flight on an airline.
Forecasting	Predicts values that can identify patterns in customer behavior.	Estimate the value of a future stream of dollar sales from a typical customer.
Sequence	Links events over time.	Identify a link between a person who buys a new house and subsequently will buy a new car within 90 days.

Text mining (mentioned in Chapter 2) is a software application used to extract key elements from unstructured data sets, discover patterns and relationships in the text materials, and summarize the information. Given that the majority of the information stored in businesses is in the form of unstructured data (e-mails, pictures, memos, transcripts, survey responses, business receipts, and so on), the need to explore and find useful information will require increased use of text mining tools in the future.

Web mining seeks to find patterns, trends, and insights into customer behavior from users of the Web. Marketers, for example, use BA services like *Google Trends* (www.google.com/trends/) and *Google Insights for Search* (http://google.about.com/od/i/g/google-insights-for-search.htm) to track the popularity of various words and phrases to learn what consumers are interested in and what they are buying.

In addition to the general software applications discussed earlier, there are focused software applications used every day by BA analysts in conducting the three steps of the BA process (see Chapter 1). These include *Microsoft Excel®* spreadsheet applications, SAS applications, and SPSS applications. *Microsoft Excel* (www.microsoft.com/) spreadsheet systems have add-in applications specifically used for BA analysis. These add-in applications broaden the use of Excel into areas of BA. *Analysis ToolPak* is an Excel add-in that contains a variety of statistical tools (for example, graphics and multiple regression) for the descriptive and predictive BA process steps. Another Excel add-in, *Solver*, contains operations research optimization tools (for example, linear programming) used in the prescriptive step of the BA process.

SAS® Analytics Pro (www.sas.com/) software provides a desktop statistical toolset allowing users to access, manipulate, analyze, and present information in visual formats. It permits users to access data from nearly any source and transform it into meaningful, usable information presented in visuals that allow decision makers to gain quick understanding of critical issues within the data. It is designed for use by analysts, researchers, statisticians, engineers, and scientists who need to explore, examine, and present data in an easily understandable way and distribute findings in a variety of formats. It is a statistical package chiefly useful in the descriptive and predictive steps of the BA process.

IBM's *SPSS software* (www-01.ibm.com/software/analytics/spss/) offers users a wide range of statistical and decision-making tools. These tools include methodologies for data collection, statistical manipulation, modeling trends in structured and unstructured data, and optimizing analytics. Depending on the statistical packages acquired, the software can cover all three steps in the BA process.

Other software applications exist to cover the prescriptive step of the BA process. One that will be used in this book is LINGO® by Lindo Systems (www.lindo.com). LINGO is a comprehensive tool designed to make building and solving optimization models faster, easier, and more efficient. LINGO provides a completely integrated package that includes an understandable language for expressing optimization models, a full-featured environment for building and editing problems, and a set of built-in solvers to handle optimization modeling in linear, nonlinear, quadratic, stochastic, and integer programming models.

In summary, the technology needed to support a BA program in any organization will entail a general information system architecture, including database management systems and progress in greater specificity down to the software that BA analysts need to compute their unique contributions to the organization. Organizations with greater BA requirements will have substantially more technology to support BA efforts, but all firms that seek to use BA as a strategy for competitive advantage will need a substantial investment in technology, because BA is a technology-dependent undertaking.

Summary

Why BA is important is directly proportional to what it costs. In this chapter, we have explored costs, but also many of the benefits of BA as a means to justify why a BA program is necessary. This chapter discussed what resources a firm would need to support a BA program. From this, three primary areas of resources were identified: personnel, data, and technology. Having identified BA personnel and needed skill sets, a review of content in BA certification exams was presented. Types of personnel specialties also were discussed. BA data internal and external sources were presented as a means of data categorization. Finally, BA technology was covered in terms of general, organization-wide information systems support to individual analyst support software packages.

In this chapter, we focused on the investment in resources needed to have a viable business analytics operation. In Chapter 4, we begin Part III, "How Can Business Analytics Be Applied?" Specifically, in the next chapter we will focus on how the resources mentioned in this chapter are placed into an organization and managed to achieve goals.

Discussion Questions

1. How does using BA certification exam content explain skill sets for BA analysts? What skill sets are necessary for BA personnel?

2. Why is leadership an important skill set for individuals looking to make a career in BA?

3. Why is categorizing data from its sources important in BA?

4. What is data quality, and why is it important in BA?

5. What is the difference between a data warehouse and a datamart?

References

Bartlett, R. (2013). *A Practitioner's Guide to Business Analytics*. McGraw-Hill, New York, NY.

Laursen, G. H. N., Thorlund, J. (2010). *Business Analytics for Managers*. John Wiley & Sons, Hoboken, NJ.

Stubbs, E. (2013). *Delivering Business Analytics*. John Wiley & Sons, Hoboken, NJ.

Stubbs, E. (2011). *The Value of Business Analytics*. John Wiley & Sons, Hoboken, NJ.

Part III

How Can Business Analytics Be Applied?

4

How Do We Align Resources to Support Business Analytics within an Organization?

Chapter objectives:

- Explain why a centralized business analytics (BA) organization structure has advantages over other structures.
- Describe the differences between BA programs, projects, and teams and how they align BA resources in firms.
- Describe reasons why BA initiatives fail.
- Describe typical BA team roles and reasons for their failures.
- Explain why establishing an information policy is important.
- Explain the advantages and disadvantages of outsourcing BA.
- Describe how data can be scrubbed.
- Explain what change management involves and what its relationship is to BA.

4.1 Organization Structures Aligning Business Analytics

According to Isson and Harriott (2013, p. 124), to successfully implement business analytics (BA) within organizations, the BA in whatever organizational form it takes must be fully integrated throughout a firm. This requires BA resources to be aligned in a way that permits a view of customer information within and across all departments, access to customer information from multiple sources (internal and external to the organization), access to historical analytics from a central repository, and alignment of technology resources so they're accountable for analytic success. The commonality of these requirements is the desire for an alignment that maximizes the flow of information into and through the BA operation, which in turn processes and shares information to desired users throughout the organization. Accomplishing this information flow objective requires consideration of differing organizational structures and managerial issues that help align BA resources to best serve an organization.

4.1.1 Organization Structures

As mentioned in Chapter 2, "Why Is Business Analytics Important?" most organizations are hierarchical, with senior managers making the strategic planning decisions, middle-level managers making tactical planning decisions, and lower-level managers making operational planning decisions. Within the hierarchy, other organizational structures exist to support the development and existence of groupings of resources like those needed for BA. These additional structures include programs, projects, and teams. A *program* in this context is the process that seeks to create an outcome and usually involves managing several related projects with the intention of improving organizational performance. A program can also be a large project. A *project* tends to deliver outcomes and can be defined as having temporary rather than permanent social systems within or across organizations to accomplish particular and clearly defined tasks, usually under time constraints. Projects are often composed of teams. A *team* consists of a group of people with skills to achieve a common purpose. Teams are especially appropriate for conducting complex tasks that have many interdependent subtasks.

The relationship of programs, projects, and teams with a business hierarchy is presented in Figure 4.1. Within this hierarchy, the organization's senior managers establish a *BA program* initiative to mandate the creation of a BA grouping within the firm as a strategic goal. A BA program does not always have an end-time limit. Middle-level managers reorganize or break down the strategic BA program goals into doable *BA project* initiatives to be undertaken in a fixed period of time. Some firms have only one project (establish a BA grouping) and others, depending on the organization structure, have multiple BA projects requiring the creation of multiple BA groupings. Projects usually have an end-time date in which to judge the successfulness of the project. The projects in some cases are further reorganized into smaller assignments, called *BA team* initiatives, to operationalize the broader strategy of the BA program. BA teams may have a long-standing time limit (for example, to exist as the main source of analytics for an entire organization) or have a fixed period (for example, to work on a specific product quality problem and then end).

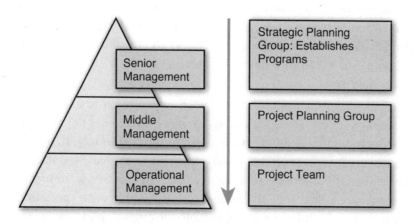

Figure 4.1 Hierarchal relationships program, project, and team planning

In summary, one way to look at the alignment of BA resources is to view it as a progression of assigned planning tasks from a BA program, to BA projects, and eventually to BA teams for implementation. As shown in Figure 4.1, this hierarchical relationship is a way to examine how firms align planning and decision-making workload to fit strategic needs and requirements.

BA organization structures usually begin with an initiative that recognizes the need to use and develop some kind of program in analytics. Fortunately, most firms today recognize this need. The question then becomes how to match the firm's needs within the organization to achieve its strategic, tactical, and operations objectives within resource limitations. Planning the BA resource allocation within the organizational structure of a firm is a starting place for the alignment of BA to best serve a firm's needs.

Aligning the BA resources requires a determination of the amount of resources a firm wants to invest. The outcome of the resource investment might identify only one individual to compute analytics for a firm. Because of the varied skill sets in information systems, statistics, and operations research methods, a more common beginning for a BA initiative is the creation of a BA team organization structure possessing a variety of analytical and management skills. (We will discuss BA teams in Section 4.1.2.) Another way of aligning BA resources within an organization is using a project structure. Most firms undertake projects, and some firms actually use a project structure for their entire organization. For example, consulting firms might view each client as a project (or product) and align their resources around the particular needs of that client. A project structure often necessitates multiple BA teams to deal with a wider variety of analytic needs. Even larger investments in BA resources might be required by firms that decide to establish a whole BA department containing all the BA resources for a particular organization. Although some firms create BA departments, the departments don't have to be large. Whatever the organization structure that is used, the role of BA is a staff (not line management) role in their advisory and consulting mission for the firm.

In general, there are different ways to structure an organization to align its BA resources to serve strategic plans. In organizations in which functional departments are structured on a strict hierarchy, separate BA departments or teams have to be allocated to each functional area, as presented in Figure 4.2. This *functional organization structure* may have the benefit of stricter functional control by the VPs of an organization and greater efficiency in focusing on just the analytics within each specialized area. On the other hand, this structure does not promote the cross-department access that is suggested as a critical success factor for the implementation of a BA program.

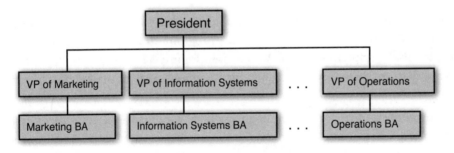

Figure 4.2 Functional organization structure with BA

The needs of each firm for BA sometimes dictate positioning BA within existing organization functional areas. Clearly, many alternative structures can house a BA grouping. For example, because BA provides information to users, BA could be included in the functional area of management information systems, with the *chief information officer* (CIO) acting as both the director of information systems (which includes database management) and the leader of the BA grouping.

An alternative organizational structure commonly found in large organizations aligns resources by project or product and is called a *matrix organization*. As illustrated in Figure 4.3, this structure allows the VPs some indirect control over their related specialists, which would include the BA specialists but also allows direct control by the project or product manager. This, similar to the functional organizational structure, does not promote the cross-department access suggested for a successful implementation of a BA program.

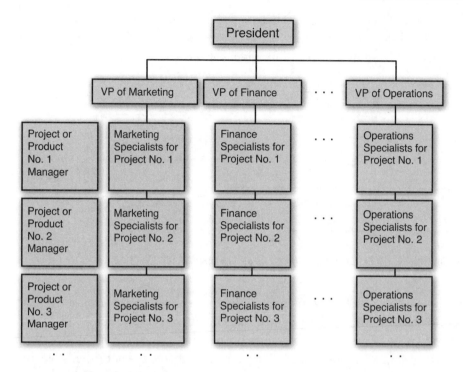

Figure 4.3 Matrix organization structure

The literature suggests that the organizational structure that best aligns BA resources is one in which a department, project, or team is formed in a staff structure where access to and from the BA grouping of resources permits access to all areas within a firm, as illustrated in Figure 4.4 (Laursen and Thorlund, 2010, pp. 191–192; Bartlett, 2013, pp. 109–111; Stubbs, 2011, p. 68). The dashed line indicates a staff (not line management) relationship. This *centralized BA organization structure* minimizes investment costs by avoiding duplications found in both the functional and the matrix styles of organization structures. At the same time, it maximizes information flow between and across functional areas in the organization. This is a logical structure for a BA group in its advisory role to the organization. Bartlett (2013, pp. 109–110) suggests other advantages of a centralized structure like the one in Figure 4.4. These include a reduction in the filtering of information traveling upward through the organization, insulation from political interests, breakdown of the *siloed functional area* communication barriers, a more central platform for reviewing important analyses that require a broader field of specialists, analytics-based group decision-making efforts, separation of the line management leadership from potential clients (for example, the VP of marketing would not necessarily come between the BA group working on customer service issues for a department within marketing), and better connectivity between BA and all personnel within the area of problem solving.

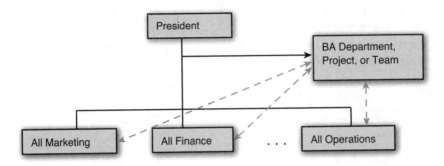

Figure 4.4 Centralized BA department, project, or team organization structure

Given the advocacy and logic recommending a centralized BA grouping, there are reasons for all BA groupings to be centralized. These reasons help explain why BA initiatives that seek to integrate and align BA resources into any type of BA group within the organization sometimes fail. The listing in Table 4.1 is not exhaustive, but it provides some of the important issues to consider in the process of structuring a BA group.

Table 4.1 Reasons for BA Initiative and Organization Failure

Reason	Description
Lack of Executive Sponsorship	Senior executive failure to recognize the value of BA eventually leads to a reduction in resources and eventual failure.
Limited Context Perception	There is an incorrect perception that analytics must be applied within a particular functional area in order to have the necessary validity to be applied to that area. Example: Financial regression analysis can only be applied correctly in the context of the finance area.
Belief of Physical Proximity	There is misperception that it takes physical proximity of the BA grouping in the business application area to be valid.
Lack of Leadership in BA Groupings	Without an advocate leader in the organization, as well as leaders in BA projects and teams to move the analysis to achieve desired goals, the entire BA effort will lead to eventual failure.
Lack of Support	Without support for needed personnel, collecting data and technology to process the data will lead to failure.
Lack of Collaboration Across All Organizational Groups	Analytics that solve problems across multiple, functional areas are more likely to be accepted and successful than those that lack the cross-over into multiple organizational groups.

Reason	Description
Lack of Skilled and Human Resources	BA departments, projects, or teams that don't have the skilled personnel to deal with the execution of analysis will eventually cause the failure of BA.
Inability to Delegate Responsibility	There is a desire to delegate responsibility to solve problems locally (a matter of trusting your own) rather than seeking help throughout the organization. This impedes the flow of problem-solving efforts by an external BA department and impedes communication of information needed to successfully apply BA.
Lack of Integrated Processes	Information that is stored in silos and not shared makes it more difficult for BA analysis to succeed.

In summary, the organizational structure that a firm may select for the positioning of its BA grouping can either be aligned within an existing organizational structure, or the BA grouping can be separate, requiring full integration within all areas of an organization. While some firms may start with a number of small teams to begin their BA program, other firms may choose to start with a full-sized BA department. Regardless of the size of the investment in BA resources, it must be aligned to allow maximum information flow between and across functional areas to achieve the most benefits BA can deliver.

4.1.2 Teams

When it comes to getting the BA job done, it tends to fall to a BA team. For firms that employ BA teams, the participants can be defined by the roles they play in the team effort. Some of the roles BA team participants undertake and their typical background are presented in Table 4.2.

Table 4.2 BA Team Participant Roles*

Title or Function	Role Description	Background or Skills of Participant
Analytics Modeler	Develop and maintain predictive and forecasting models to provide insight.	Statistics, operation research, analytic modeling.
Analytics Process Designer	Develop and enforce reusable processes to reduce BA execution time.	Management consultant, process mapping, systems design.
Analytics Analyst	Respond to BA inquiries from functional areas within the firm to gain insight.	Reporting, problem solving, communicating, and providing customer service.
BA Team Head	Provide leadership to BA team, define strategies and tactics to ensure improved business performance, and interface with management.	BA manager or administrator.

Title or Function	Role Description	Background or Skills of Participant
Business Domain Expert	Provide business experience to ensure relevance of insight, help interpret business measures and the meaning of data.	Business experience in the area of the problem or opportunity.
Data Manager	Ensure data availability and access while minimizing costs.	Data modeling or warehousing, experience in data quality processes.
Implementation Specialist	Ensure rapid and robust model deployment to reduce time in interface.	Information system and data warehousing expertise, enterprise architecture experience.
Monitoring Analyst	Identify, establish, and enforce common analytics to be used to measure value and optimize effort.	Management and BA expert, predictive and financial modeling, process design, and team mentoring.

Source: Adapted from Stubbs (2013), pp.137–149; Stubbs (2011), Table 3.3; Laursen and Thorlund (2010), p.15.

Aligning BA teams to achieve their tasks requires collaboration efforts from team members and from their organizations. Like BA teams, *collaboration* involves working with people to achieve a shared and explicit set of goals consistent with their mission. BA teams also have a specific mission to complete. Collaboration through teamwork is the means to accomplish their mission.

Team members' need for collaboration is motivated by changes in the nature of work (no more silos to hide behind, much more open environment, and so on), growth in professions (for example, interactive jobs tend to be more professional, requiring greater variety in expertise sharing), and the need to nurture innovation (creativity and innovation are fostered by collaboration with a variety of people sharing ideas). To keep their job and to progress in any business career, particularly in BA, team members must encourage working with other members inside a team and out. For organizations, collaboration is motivated by the changing nature of information flow (that is, hierarchical flows tend to be downward, whereas in modern organizations, flow is in all directions) and changes in the scope of business operations (that is, going from domestic to global allows for a greater flow of ideas and information from multiple sources in multiple locations).

How does a firm change its culture of work and business operations to encourage collaboration? One way to affect the culture is to provide the technology to support a more open, cross-departmental information flow. This includes e-mail, instant messaging, *wikis* (collaboratively edited works, like Wikipedia), use of social media and networking through *Facebook* and *Twitter*, and encouragement of activities like collaborative writing, reviewing, and editing efforts. Other technology supporting collaboration includes webinars, audio and video conferencing, and even the use of iPads to enhance face-to-face communication. These can be tools that change the culture of a firm to be more open and communicative.

Reward systems should be put into place to acknowledge team effort. Teams should be commended for their performance, and individuals should be praised for performance in a team. While middle-level managers build teams, coordinate their work, and monitor their performance, senior management should establish collaboration and teamwork as a vital function.

Despite the collaboration and best of intentions, BA teams sometimes fail. There are many reasons for this, but knowing some of the more common ones can help managers avoid them. Some of the more common reasons for team failure are presented in Table 4.3. They also represent issues that can cause a BA program to become unaligned and unproductive.

Table 4.3 Reasons for BA Team Failures*

Reason for Failure	Descriptions
Lack of Communication	It is not enough to come up with valuable information for decision-making and to find business opportunities in data. That information must be shared with users, clients, and everyone within a firm for benefit to come from it. It is only when analytics show a tangible and beneficial outcome that they are considered BA. If those results are not communicated on a continual basis, BA teams can be perceived to provide less value to the organization.
Failure to Deliver	Not every BA team will be able to deliver valued information if the team lacks the ability or resources to deliver needed answers and information. The greater the number of BA team failures, the greater are the chances that the team will be eliminated.
Lack of Justification	BA teams require resource allocations. Those allocations come from other departments that supposedly benefit from BA contributions. Without the role of BA and its potential contributions to a firm being clearly spelled out, users might not associate the ongoing efforts of a BA team as being worth the money spent on them.
Fail to Provide Value	BA teams have to sell their roles and suggested solutions or ideas. Without a clear understanding of value for potential users, the team faces a hard sell.
Inability to Prove Success	BA teams need to document and measure the impact of their ideas and suggestions. Without that proof, potential users might not support future BA efforts.

*****Source**: Adapted from Flynn (2008), pp. 99–106, and Stubbs (2011), p. 89.

4.2 Management Issues

Aligning organizational resources is a management function. There are general management issues that are related to a BA program, and some are specifically important to operating a BA department, project, or team. The ones covered in this section include establishing an information policy, outsourcing business analytics, ensuring data quality, measuring business analytics contribution, and managing change.

4.2.1 Establishing an Information Policy

There is a need to manage information. This is accomplished by establishing an *information policy* to structure rules on how information and data are to be organized and maintained and who is allowed to view the data or change it. The information policy specifies organizational rules for sharing, disseminating, acquiring, standardizing, classifying, and inventorying all types of information and data. It defines the specific procedures and accountabilities that identify which users and organizational units can share information, where the information can be distributed, and who is responsible for updating and maintaining the information.

In small firms, business owners might establish the information policy. For larger firms, *data administration* may be responsible for the specific policies and procedures for data management (Siegel and Shim, 2003, p. 280). Responsibilities could include developing the information policy, planning data collection and storage, overseeing database design, developing the data dictionary, as well as monitoring how information systems specialists and end user groups use data.

A more popular term for many of the activities of data administration is *data governance*, which includes establishing policies and processes for managing the availability, usability, integrity, and security of the data employed in businesses. It is specifically focused on promoting data privacy, data security, data quality, and compliance with government regulations.

Such information policy, data administration, and data governance must be in place to guard and ensure data is managed for the betterment of the entire organization. These steps are also important in the creation of database management systems (see Chapter 3, "What Resource Considerations Are Important to Support Business Analytics?") and their support of BA tasks.

4.2.2 Outsourcing Business Analytics

Outsourcing can be defined as a strategy by which an organization chooses to allocate some business activities and responsibilities from an internal source to an external source (Schniederjans et al., 2005, pp. 3–4). Outsourcing business operations is a strategy that an organization can use to implement a BA program, run BA projects, and operate BA teams. Any business activity can be outsourced, including BA. Outsourcing is an important BA management activity that should be considered as a viable alternative in planning an investment in any BA program.

BA is a staff function that is easier to outsource than other line management tasks, such as running a warehouse. To determine if outsourcing is a useful option in BA programs, management needs to balance the advantages of outsourcing with its disadvantages. Some of the advantages of outsourcing BA include those listed in Table 4.4.

Table 4.4 Advantages of Outsourcing BA

Advantage of Outsourcing BA	Description
Less Expensive	Maintaining a fully functioning BA department when analytics might only be useful periodically may be more expensive than occasionally hiring an outside consulting BA firm to solve a problem.
Superior Analytics	The pool of analytic capabilities is always going to be greater outside a firm.
More Staffing Flexibility	Staff positions are often the first cut in economy downturns. Using consultants is easier and less expensive than hiring full-time BA staff. Outsourcing permits more flexibility to add and reduce BA services as needed.
New Knowledge	Experienced BA consultants bring a variety of knowledge and experience from having worked with many other firms. That type of experience may be of great competitive advantage.

Nevertheless, there are disadvantages of outsourcing BA. Some of the disadvantages to outsourcing are presented in Table 4.5.

Table 4.5 Disadvantages of Outsourcing BA

Disadvantages of Outsourcing BA	Description
Loss of Control	Once outsourced, most of the control of a BA project is under the control of the outsourcing firm. The client firm might lose not only control, but also opportunities for new and unique information, which the outsourcing firm may not share with the client.

Disadvantages of Outsourcing BA	Description
Difficulties in Managing the Relationship	Client firms may find it difficult to communicate with the outsourcing firm because of distance, differing culture, language issues, and more. The lack of management could cause substantial problems with customer service and product quality.
Weakens Innovation	Having outsourced a client firm's internal experts, the remaining collaboration within the firm's personnel is reduced, and that reduces the opportunity for innovation efforts through shared collaboration.
Risk of Information	Outsourcing staff are exposed to client proprietary information, including innovations in analytics. This information could be shared with other competing firms, placing the client firm at risk.
Worthless Analytics	Sometimes outsourcing partners are less capable than internal analysts, wasting time and money.

Managing outsourcing of BA does not have to involve the entire department. Most firms outsource projects or tasks found to be too costly to assign internally. For example, firms outsource cloud computing services to outside vendors (Laudon and Laudon, 2012, p. 511), and other firms outsource software development or maintenance of legacy programs to offshore firms in low-wage areas of the world to cut costs (Laudon and Laudon, 2012, p. 192).

Outsourcing BA can also be used as a strategy to bring BA into an organization (Schniederjans et al., 2005, pp. 24–27). Initially, to learn how to operate a BA program, project, or team, an outsource firm can be hired for a limited, contracted period. The client firm can then learn from the outsourcing firm's experience and instruction. Once the outsourcing contract is over, the client firm can form its own BA department, project, or team.

4.2.3 Ensuring Data Quality

Business analytics, if relevant, is based on data assumed to be of high quality. *Data quality* refers to accuracy, precision, and completeness of data. High-quality data is considered to correctly reflect the real world in which it is extracted. Poor-quality data caused by data entry errors, poorly maintained databases, out-of-date data, and incomplete data usually leads to bad decisions and undermines BA within a firm. Organizationally, the database management systems (DBMS, mentioned in Chapter 3) personnel are managerially responsible for ensuring data quality. Because of its importance and the possible location of the BA department outside the management information systems department (which usually hosts the DBMS), it is imperative that whoever leads the BA program should seek to ensure data quality efforts are undertaken.

Ideally, a properly designed database with organization-wide data standards and efforts taken to avoid duplication or inconsistent date elements should have high-quality data. Unfortunately, times are changing, and more organizations allow customers and suppliers to enter data into databases directly via the Web. As a result, most of the quality problems originate from data input such as misspelled names, transposed numbers, or incorrect or missing codes.

An organization needs to identify and correct faulty data and establish routines and procedures for editing data in the database. The analysis of data quality can begin with a *data quality audit*, in which a structured survey or inspection of accuracy and level of completeness of data is undertaken. This audit may be of the entire database, just a sample of files, or a survey of end users for perceptions of the data quality. If during the data quality audit files are found that have errors, a process called *data cleansing* or *data scrubbing* is undertaken to eliminate or repair data. Some of the areas in a data file that should be inspected in the audit and suggestions on how to correct them are presented in Table 4.6.

Table 4.6 Quality Data Inspection Items and Recommendations

Data Inspection Items	Description and Cleansing/Scrubbing Recommendation
Current Data	Check to make sure the data is current. If it is out of date, remove it.
Completeness	Check to see if there is missing data. If more than 50 percent is missing, remove the entire file from the database.
Relevance	Check to see if the data is no longer relevant for the purpose for which it was collected. If it's no longer relevant, consider removing it from the database.
Duplication	Check to see if duplicate data files exist in the database. Remove duplicate data.
Outliers	Check for extreme values (outliers) in quantitative data files for possible errors in data coding. Remove from the data file any suspected of being in error, or repair the data.
Inconsistent Values	If data fields contain both characters and real numbers data where only characters or numbers should be, explore repairing the data.
Coding	If suspicious or unknown coding of data exists in data files, remove from the database or repair the coding of data.

4.2.4 Measuring Business Analytics Contribution

The investment in BA must continually be justified by communicating the BA contribution to the organization for ongoing projects. This means that performance analytics should be computed for every BA project and BA team initiative. These analytics should provide an estimate of the tangible and intangible values being delivered to the organization. This should also involve establishing a communication strategy to promote the value being estimated.

Measuring the value and contributions that BA brings to an organization is essential to helping the firm understand why the application of BA is worth the investment. Some BA contribution estimates can be computed using standard financial methods, such as *payback period* (how long it takes for the initial costs to be returned by profit) or return on investment (ROI) (see Schniederjans et al., 2010, pp. 90–132), where dollar values or quantitative analysis is possible. When intangible contributions are a major part of the contribution being delivered to the firm, other methods like cost/benefit analysis (see Schniederjans et al., 2010, pp. 143–158), which include intangible benefits, should be used.

The continued measurement of value that BA brings to a firm is not meant to be self-serving, but it aids the organization in aligning efforts to solve problems and find new business opportunities. By continually running BA initiatives, a firm is more likely to identify internal activities that should and can be enhanced by employing optimization methodologies during the Prescriptive step of the BA process introduced in Chapter 1, "What Is Business Analytics?" It can also help identify underperforming assets. In addition, keeping track of investment payoffs for BA initiatives can identify areas in the organization that should have a higher priority for analysis. Indeed, past applications and allocations of BA resources that have shown significant contributions can justify priorities established by the BA leadership about where there should be allocated analysis efforts within the firm. They can also help acquire increases in data support, staff hiring, and further investments in BA technology.

4.2.5 Managing Change

Wells (2000) found that what is critical in changing organizations is organizational culture and the use of change management. *Organizational culture* is how an organization supports cooperation, coordination, and empowerment of employees (Schermerhorn, 2001, p. 38). *Change management* is defined as an approach for transitioning the organization (individuals, teams, projects, departments) to a changed and desired future state (Laudon and Laudon, 2012, pp. 540–542). Change management is a means of implementing change in an organization, such as adding a BA department (Schermerhorn, 2001, pp. 382–390). Changes in an organization can be either planned (a result of specific and planned efforts at change with direction by a change leader) or unplanned (spontaneous changes without direction of a change leader). The application of BA invariably will result in both types of changes because of BA's specific problem-solving role (a desired, planned change to solve a problem) and opportunity-finding exploratory nature (unplanned new knowledge opportunity changes) of BA. Change management can also target almost everything that makes up an organization (see Table 4.7).

Table 4.7 Change Management Targets*

Change Target	Description
Culture	This represents the changing values and norms of the individuals and groups that make up the organization. BA has to sell itself in some situations, build trust, and alter decision-making. It often requires a different culture of thinking about decision-making.
Organization Structure	This is the changing organizational lines of authority and communication. The cross-departmental nature of BA positions may provide information that changes the organization and alters relationships and tasks.
Personnel	BA information about the need for human resource changes in attitudes and skills can mandate changes that permit an organization to achieve higher business performance levels.
Tasks	BA analysis might find that some job designs, specifications, and descriptions that employees perform need to have their objectives and goals changed to achieve higher business performance levels.
Technology	BA analysis might find information system technology used in the design and workflow that integrate employees and equipment into operating systems and require change to achieve higher business performance levels.

Source: Adapted from Figure 7 in Schniederjans and Cao (2002), pp. 261.

It is not possible to gain the benefits of BA without change. The intent is change that involves finding new and unique information on which change should take place in people, technology systems, or business conduct. By instituting the concept of change management within an organization, a firm can align resources and processes to more readily accept changes that BA may suggest. Instituting the concept of change management in any firm depends on the unique characteristics of that firm. There are, though, a number of activities in common with successful change management programs, and they apply equally to changes in BA departments, projects, or teams. Some of these activities that lead to change management success are presented as best practices in Table 4.8.

Table 4.8 Change Management Best Practices

Best Practice	Description
Champion	Change is scary business for some, and a strong leader for change can champion the change effort, calming fears and explaining the need for change. The champion also helps direct efforts, motivate change, and keep the change activities on track.
Clearly Stated Goals	Any type of change should be clearly defined, including what the changes are, which personnel have to change, and what the processes involve and how they affect technology. This would also include deadlines needed to keep the change effort on track.
Good Communication	To avoid resistance to change (a natural norm to anything that is new), it is useful to help those facing the change understand its value through effective and repeated communications, keeping them informed on progress and easing fears.
Measured Performance	Any goals stated prior to the launch of change can be used to measure performance during the changeover period. Seeing business performance improve with changes can motivate further change and support by those impacted.
Senior Management Support	Critical to all BA departments, projects, or teams is the need for senior management to support change efforts. Sometimes that support is in direct dollars, and sometimes it's in lending authority to get resources needed for BA work.

Summary

Structuring a BA department, undertaking a BA project, or setting up a BA team within an organization can largely determine successfulness in aligning resources to achieve information-sharing goals. In this chapter, several organization structures (functional, matrix, and centralized) were discussed as possible homes for BA resource groupings. The role of BA teams as an important organizational resource aligning tool was also presented. In addition, this chapter discussed reasons for BA organization and team failures. Other managerial issues included in this chapter were establishing an information policy, outsourcing business analytics, ensuring data quality, measuring business analytics contribution, and managing change.

Once a firm has set up the internal organization for a BA department, program, or project, the next step is to undertake BA. In the next chapter, we begin the first of the three chapters devoted to detailing how to undertake the three steps of the BA process.

Discussion Questions

1. The literature in management information systems consistently suggests that a decentralized approach to resource allocation is the most efficient. Why then do you think the literature in BA suggests that the opposite—a centralized organization—is the best structure?

2. Why is collaboration important to BA?

3. Why is organization culture important to BA?

4. How does establishing an information policy affect BA?

5. Under what circumstances is outsourcing BA good for the development of BA in an organization?

6. Why do we have to measure BA contributions to an organization?

7. How does data quality affect BA?

8. What role does change management play in BA?

References

Bartlett, R. (2013). *A Practitioner's Guide to Business Analytics*. McGraw-Hill, New York, NY.

Flynn, A. E. (2008). *Leadership in Supply Management*. Institute for Supply Management, Inc., Tempe, AZ.

Isson, J. P., Harriott, J. S. (2013). *Win with Advanced Business Analytics*. John Wiley & Sons, Hoboken, NJ.

Laudon, K., Laudon, J. (2012). *Essentials of Management Information Systems*, 10th Ed. Prentice Hall, Upper Saddle River, NJ.

Laursen, G. H. N., Thorlund, J. (2010). *Business Analytics for Managers*. John Wiley & Sons, Hoboken, NJ.

Schermerhorn, J. R. (2001). *Management*, 6th ed., John Wiley & Sons, New York, NY.

Schniederjans, M. J., Cao, Q. (2002). *E-commerce Operations Management*. World Scientific, Singapore.

Schniederjans, M. J., Hamaker, H. L., Schniederjans, A. M. (2010). *Information Technology Investment*. World Scientific, Singapore.

Schniederjans, M. J., Schniederjans, A. M., Schniederjans, D. G. (2005). *Outsourcing and Insourcing in an International Context*. M. E. Sharpe, Armonk, NY.

Siegel, J., Shim, J. (2003). *Database Management Systems*. Thomson/South-Western, Mason, OH.

Stubbs, E. (2013). *Delivering Business Analytics*. John Wiley & Sons, Hoboken, NJ.

Stubbs, E. (2011). *The Value of Business Analytics*. John Wiley & Sons, Hoboken, NJ.

Wells, M. G. (2000). "Business Process Re-Engineering Implementations Using Internet Technology." *Business Process Management Journal*, Vol. 6, No. 2, 2000, pp. 164–184.

5

What Is Descriptive Analytics?

Chapter objectives:

- Explain why we need to visualize and explore data.
- Describe statistical charts and how to apply them.
- Describe descriptive statistics useful in the descriptive business analytics (BA) process.
- Describe sampling methods useful in BA and where to apply them.
- Describe what sampling estimation is and how it can aid in the BA process.
- Describe the use of confidence intervals and probability distributions.
- Explain how to undertake the descriptive analytics step in the BA process.

5.1 Introduction

In any BA undertaking, referred to as *BA initiatives* or *projects*, a set of objectives is articulated. These objectives are a means to align the BA activities to support strategic goals. The objectives might be to seek out and find new business opportunities, to solve operational problems the firm is experiencing, or to grow the organization. It is from the objectives that exploration via BA originates and is in part guided. The directives that come down, from the strategic planners in an organization to the BA department or analyst, focus the tactical effort of the BA initiative or project. Maybe the assignment will be one of exploring internal marketing data for a new marketing product. Maybe the BA assignment will be focused on enhancing service quality by collecting engineering and customer service information. Regardless of the type of BA assignment, the first step is one of exploring data and revealing new, unique, and relevant information to help the organization advance its goals. Doing this requires an exploration of data.

This chapter focuses on how to undertake the first step in the BA process: *descriptive analytics*. The focus in this chapter is to acquaint readers with more common descriptive analytic tools used in this step and available in SAS software. The treatment here is not computational but informational regarding the use and meanings of these analytic tools in support of BA. For purposes of illustration, we will use the data set in Figure 5.1 representing four different types of product sales (Sales 1, Sales 2, Sales 3, and Sales 4).

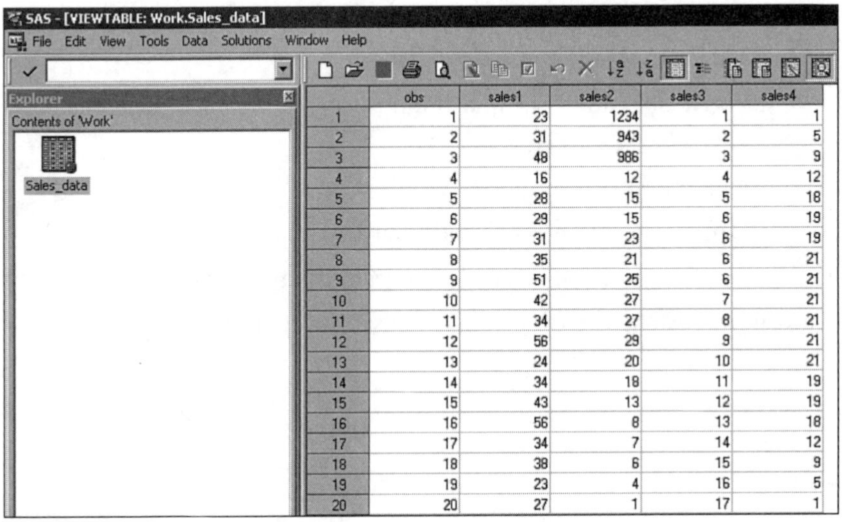

	obs	sales1	sales2	sales3	sales4
1	1	23	1234	1	1
2	2	31	943	2	5
3	3	48	986	3	9
4	4	16	12	4	12
5	5	28	15	5	18
6	6	29	15	6	19
7	7	31	23	6	19
8	8	35	21	6	21
9	9	51	25	6	21
10	10	42	27	7	21
11	11	34	27	8	21
12	12	56	29	9	21
13	13	24	20	10	21
14	14	34	18	11	19
15	15	43	13	12	19
16	16	56	8	13	18
17	17	34	7	14	12
18	18	38	6	15	9
19	19	23	4	16	5
20	20	27	1	17	1

Figure 5.1 Illustrative sales data sets

When using SAS, data sets are placed into files like that in Figure 5.2.

Creating the data set in Figure 5.2 requires a sequence of SAS steps. Because this is the first use of SAS, the sequence of instructions is illustrated in Figures 5.3 through 5.9. These steps should be familiar to experienced SAS users. Depending on the SAS version and intent of the data file, the images in these figures may be slightly different.

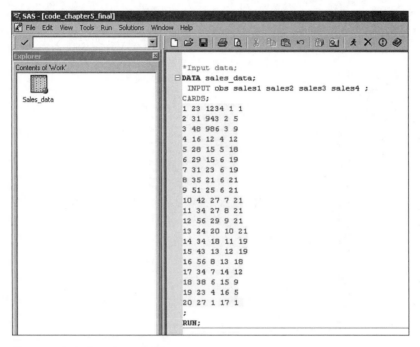

```
                          *Input data;
                          DATA sales_data;
                            INPUT obs sales1 sales2 sales3 sales4 ;
                          CARDS;
                          1 23 1234 1 1
                          2 31 943 2 5
                          3 48 986 3 9
                          4 16 12 4 12
                          5 28 15 5 18
                          6 29 15 6 19
                          7 31 23 6 19
                          8 35 21 6 21
                          9 51 25 6 21
                          10 42 27 7 21
                          11 34 27 8 21
                          12 56 29 9 21
                          13 24 20 10 21
                          14 34 18 11 19
                          15 43 13 12 19
                          16 56 8 13 18
                          17 34 7 14 12
                          18 38 6 15 9
                          19 23 4 16 5
                          20 27 1 17 1
                          ;
                          RUN;
```

Figure 5.2 SAS coding of sales data set

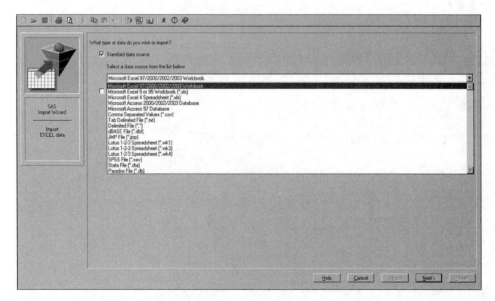

Figure 5.3 Creating a workbook file for the data set

	A	B	C	D
1	Sales 1	Sales 2	Sales 3	Sales 4
2	23	1234	1	1
3	31	943	2	5
4	48	896	3	9
5	16	12	4	12
6	28	15	5	18
7	29	15	6	19
8	31	23	6	19
9	35	21	6	21
10	51	25	6	21
11	42	27	7	21
12	34	27	8	21
13	56	29	9	21
14	24	20	10	21
15	34	18	11	19
16	43	13	12	19
17	56	8	13	18
18	34	7	14	12
19	38	6	15	9
20	23	4	16	5
21	27	1	17	1

Figure 5.4 Excel data file used in the creation of the SAS data file

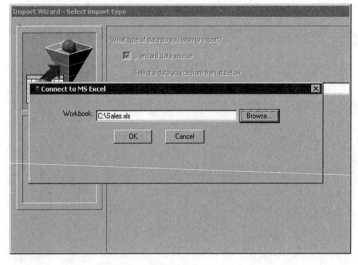

Figure 5.5 Step to pull the data set from an Excel (or any) file using SAS

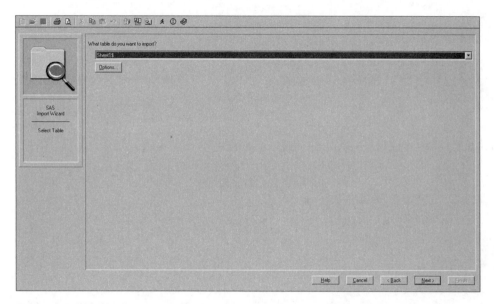

Figure 5.6 Identify file using SAS

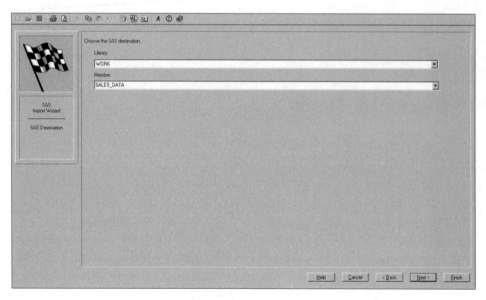

Figure 5.7 Label SAS file as SALES_DATA

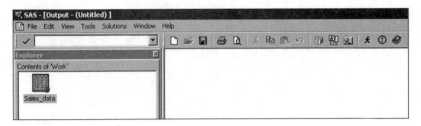

Figure 5.8 Shows the SALES_DATA SAS file is created

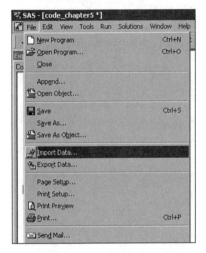

Figure 5.9 Step to import the created file

SAS permits the use of many different sources for data sets or data files to be entered into an SAS program. Big data files in Excel, SPSS, or other software applications can be brought into SAS programs using a similar set of steps presented in this section. Once the data sets are structured for use with SAS, there is still considerable SAS programming effort needed to glean useful information from any big data or small data files. Fortunately, SAS provides the means by which any sized data set can be explored and visualized by BA analysts. Because SAS is a programming language, it permits a higher degree of application customization than most other statistical software.

5.2 Visualizing and Exploring Data

There is no single best way to explore a data set, but some way of conceptualizing what the data set looks like is needed for this step of the BA process. Charting is often employed to visualize what the data might reveal.

When determining the software options to generate charts in SAS, consider that the software can draft a variety of charts for the selected variables in the data sets. Using the data in Figure 5.1, charts can be created for the illustrative sales data sets. Some of these charts are discussed in Table 5.1 as a set of exploratory tools that are helpful in understanding the informational value of data sets. The chart to select depends on the objectives set for the chart. The SAS program statements used to create each chart in Table 5.1 are provided.

Table 5.1 Statistical Charts Useful in BA

Type of Chart	Application Notes	Chart Example
| Bar | • Can be horizontal, vertical, cone, or cyclically shaped and multi-dimensional with overlaying variables.<br><br>• Ideal for showing comparative improvement over time.<br><br>• Example: Bars showing productivity of one person versus another. | ```
proc gchart data=sales_data;
hbar obs /sumvar=sales2
midpoints=1 to 20 by 1;
run;
quit;
```<br><br> |
```

Type of Chart	Application Notes	Chart Example
Column	• Same as a bar chart.	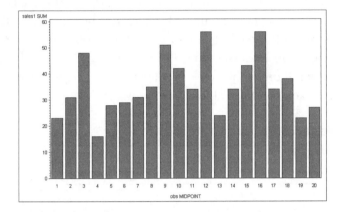
Line	• Ideal for showing linear trend and other linear or nonlinear appearance. • Best applied with time series data with time as the X-axis.	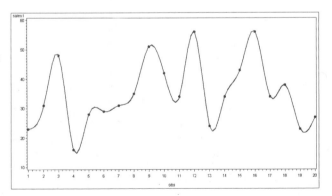

For the Column chart:

```
proc gchart data=sales_data;
vbar obs /sumvar=sales1
midpoints=1 to 20 by 1;
run;
quit;
```

For the Line chart:

```
symbol value=dot interpol=sms line=1 width=2;
proc gplot data=sales_data;
plot sales1*obs;
run;
quit;
```

Type of Chart	Application Notes	Chart Example
Pie	<ul><li>Useful in conceptualizing proportions.</li><li>Various other versions, like the donut chart (with a hollow center), can also be used.</li><li>Useful when the number of variables is limited (not like the illustration to the right).</li></ul>	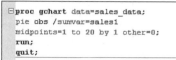 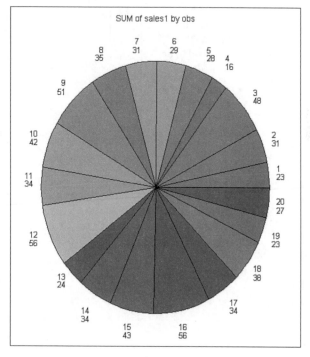

Type of Chart	Application Notes	Chart Example

Scatter	• Useful when patterns are observed in the data sets.
	
	• Useful when outliers are observed in the data that may need to be cleaned out.
	
	• Outline trends that a linear chart can augment.
		``
		proc sgplot data=sales_data;
		  scatter x=obs y=sales4;
		run;
		``

Type of Chart	Application Notes	Chart Example
Histogram	• Ideal to help reveal frequency distributions in variable data sets. • Reduces the size of data by grouping data points into frequencies.	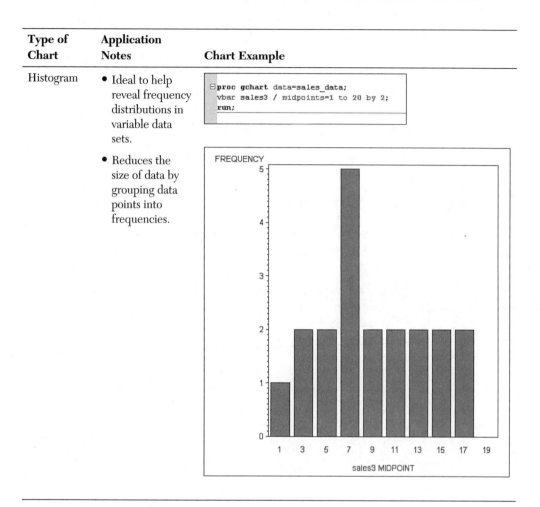

The charts presented in Table 5.1 reveal interesting facts. The column chart is useful in revealing the almost perfect linear trend in the Sales 3 data, whereas the scatter chart reveals an almost perfect nonlinear function in Sales 4 data. Additionally, the cluttered pie chart with 20 different percentages illustrates that all charts can or should be used in some situations. The best practices suggest charting should be viewed as an exploratory activity of BA. BA analysts should run a variety of charts and see which ones reveal interesting and useful information. Those charts can be further refined to drill down to more detailed information and more appropriate charts related to the objectives of the BA initiative.

Of course, a cursory review of the Sales 4 data in Figure 5.1 makes the concave appearance of the data in the scatter chart in Table 5.1 unnecessary. But most BA problems involve big data—so large as to make it impossible to just view it and make judgment calls on structure or appearance. This is why descriptive statistics can be employed to view the data in a parameter-based way in the hopes of better understanding the information that the data has to reveal.

5.3 Descriptive Statistics

SAS has a number of useful statistics that can be automatically computed for the variables in the data sets. The SAS printout of the sales data from Figure 5.1 is summarized in Table 5.2. Some of these descriptive statistics are discussed in Table 5.3 as exploratory tools that are helpful in understanding the informational value of data sets.

Table 5.2 SAS Descriptive Statistics

	N	Range	Min	Max	Sum	Mean		Std. Dev.	Variance	Skewedness		Kurtosis	
	Statistic	Statistic	Statistic	Statistic	Statistic	Statistic	Std. Error	Statistic	Statistic	Statistic	Std. Error	Statistic	Std. Error
Sales 1	20	40	16	56	703	35.15	2.504	11.198	125.397	.490	.512	-.429	.992
Sales 2	20	1233	1	1234	3344	167.20	83.686	374.254	140065.853	2.241	.512	3.636	.992
Sales 3	20	16	1	17	171	8.55	1.065	4.763	22.682	.272	.512	-.988	.992
Sales 4	20	20	1	21	292	14.60	1.603	7.170	51.411	-.824	.512	-.825	.992
Valid N	20												

Table 5.3 Descriptive Statistics Useful in BA

Statistics	Computation (in Data Set)	Application Area	Example	Application Notes
N or Count	Number of values.	Any.	Sample size of a company's transactions during a month.	Useful in knowing how many items were used in the statistics computations.
Sum	Total of the values in the entire data set.	Any.	Total sales for a company.	Useful in knowing the total value.
Mean	Average of all values.	Any.	Average sales per month.	Useful in capturing the central tendency of the data set.
Median	Midpoint value in the data set arranged from high to low.	Finding the midpoint in the distribution of data.	Total income for citizens of a country.	Useful in finding the point where 50 percent of the data is above and below.
Mode	Most common value in the data set.	Where values are highly repeated in the data set.	Fixed annual salaries where a limited number of wage levels is used.	Useful in declaring a common value in highly repetitive data sets.
Maximum/ Minimum	Largest and smallest values, respectively.	To conceptualize the spread of the data's distribution.	Largest and smallest sales in a day.	Useful in providing a scope or end points in the data.
Range	Difference between the max and min values.	A crude estimate of the spread of the data's distribution.	Spread of sales in units during a month.	Useful as a simple estimate of dispersion.
Standard deviation	Square root of the average of the differences squared between the mean and all other values in the data set.	A precise estimate of the spread of the data's distribution from a mean value in terms of the units used in its computation.	Standard deviation in dollars from mean sales.	The smaller the value, the less the variation and the more predictable using the data set.

Statistics	Computation (in Data Set)	Application Area	Example	Application Notes
Variance	Average differences squared between the mean and all other values.	A variance estimate of the spread of the data's distribution from a mean value, not in terms of the units used in its computation.	Measure of variance that is best used when compared with another variance computed on the same data set.	The smaller the value, the less the variation and the more predictable the data set.
(Coefficient of) Skewedness	Positive or negative values. If value sign is +, distribution is positively skewed; if −, it is negatively skewed. The larger the value, the greater it is skewed.	Measure of the degree of asymmetry of data about a mean.	As the age of residents in a country becomes older, the population age distribution becomes more negatively skewed.	The closer the value is to 0, the better the symmetry. A positively skewed distribution has its largest allocation to the left, and a negative distribution to the right.
(Coefficient of) Kurtosis	Value where less than 3 means a flat distribution and more than 3 means a peaked distribution.	Measure of the degree of spread vertically in a distribution about a mean. Also, it reveals a positive and a negative symmetry depending on its sign.	Distribution of customers at lunch and dinner times peaks and then flattens out.	The closer the value is to 2, the less is the kurtosis (peaking or flattening in the distribution).
Standard Error (of the Mean)	Mean of the sample standard deviation (that is, a standard deviation adjusted to reflect a sample size).	Standard deviation of a sampling distribution.	Standard deviation in dollars from mean sales based on a sample.	The smaller the value, the less the variation and the more predictable the sample data set.
Sample Variance	Same as variance but adjusted for sample sizes.	Variance estimate of the spread of the sampling data distribution.	Measure of variance when sampling is used for collection purposes.	The smaller the value, the less the variation and the more predictable the sample data set.

Fortunately, we do not need to compute these statistics to know how to use them. Computer software provides these descriptive statistics when they're needed or requested. When you look at the data sets for the four variables in Figure 5.1 and at the statistics in Table 5.2, there are some obvious conclusions based on the detailed statistics from the data sets. It should be no surprise that Sales 2, with a few of the largest values and mostly smaller ones making up the data set, would have the largest variance statistics (standard deviation, sample variance, range, maximum/minimum). Also, Sales 2 is highly, positively skewed (Skewedness > 1) and highly peaked (Kurtosis > 3). Note the similarity of the mean, median, and mode in Sales 1 and the dissimilarity in Sales 2. These descriptive statistics provide a more precise basis to envision the behavior of the data. Referred to as *measures of central tendency,* the mean, median, and mode can also be used to clearly define the direction of a skewed distribution. A negatively skewed distribution orders these measures such that mean<median<mode, and a positive skewed distribution orders them such that mode<median<mean.

So what can be learned from these statistics? There are many observations that can be drawn from this data. Keep in mind that, in dealing with the big data sets, one would only have the charts and the statistics to serve as a guide in determining what the data looks like. Yet, from these statistics, one can begin describing the data set. So in the case of Sales 2, it can be predicted that the data set is positively skewed and peaked. Note in Figure 5.10 that the histogram of Sales 2 is presented. The SAS chart also overlays a normal distribution (a bell-shaped curve) to reflect the positioning of the mean (highest point on the curve, 167.2) and the way the data appears to fit the normal distribution (not very well in this situation). As expected, the distribution is positively distributed with a substantial variance between the large values in the data set and the many more smaller valued data points.

We also know that substantial variance in the data points making up the data set is highly diverse—so much so that it would be difficult to use this variable to predict future behavior or trends. This type of information may be useful in the further steps of the BA process as a means of weeding out data that will not help predict anything useful. Therefore, it would not help an organization improve its operations.

Sometimes big data files become so large that certain statistical software systems cannot manipulate them. In these instances, a smaller but representative sample of the data can be obtained if necessary. Obtaining the sample for accurate prediction of business behavior requires understanding the sampling process and estimation from that process.

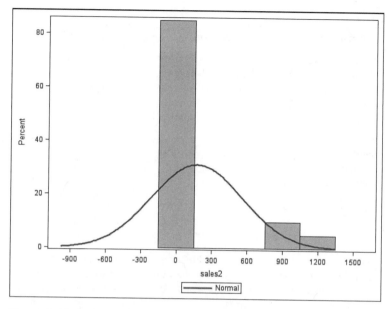

Figure 5.10 SAS histogram of Sales 2 data

5.4 Sampling and Estimation

The estimation of most business analytics requires sample data. In this section, we discuss various types of sampling methods and follow that with a discussion of how the samples are used in sampling estimation.

5.4.1 Sampling Methods

Sampling is an important strategy of handling large data. Big data can be cumbersome to work with, but a smaller sample of items from the big data file can provide a new data file that seeks to accurately represent the population from which it comes. In sampling data, there are three components to recognize: a population, a sample, and a *sample element* (the items that make up the sample). A firm's collection of customer service performance documents for one year could be designated as a population of customer service performance for that year. From that population, a sample of a lesser number of sample elements (the individual customer service documents) can be drawn to reduce the effort of working with the larger data. Several sampling methods can be used to arrive at a representative sample. Some of these sampling methods are presented in Table 5.4.

Table 5.4 Sampling Methods

Sampling Method	Description	Application	Application Notes
Simple Random	Allows each sample element in a population to have an equal chance of selection.	Selecting customers based on their percentage of occurrence as a member of a particular race.	Sample size must be sufficient to avoid sampling bias.
Systematic Random (or Period)	Selects sample elements from a population based on a fixed number in an interval.	Selecting every fifth person leaving an airport to interview.	Assumes the sample elements order in the interval is presented in a random fashion; otherwise, it can result in sampling bias.
Stratified Random	Stage 1: Divide a population into groups (called strata); Stage 2: Apply simple random sampling.	Randomly selecting an equal number of people in each of three different economic strata.	Strata must be representative of the population, or it can result in sampling bias.
Cluster Random	Stage 1: Group sample elements geographically (called clusters); Stage 2: Apply simple random sampling.	Randomly selecting an equal number of people from voting districts.	Cluster must be representative of the population, or it can result in sampling bias.
Quota	Based on a fixed quota or number of sample elements.	Selecting the first 200 people who enter a store.	• Mainly used to save time and money. • Sample size must be sufficient to avoid sampling bias.
Judgment	Selects sample elements based on expert judgment.	Selecting candidates for an interview with a special offer based on their appearance.	Prone to bias without defined criteria for selection because of dependency on interviewer experience.

The simple, systematic, stratified, and cluster random methods are based on some kind of probability of their occurrence in a population. The quota and judgment methods are nonprobability-based tools. Although the randomization process in some methods helps ensure representative samples being drawn from the population, sometimes because of cost or time constraints, nonprobability methods are the best choice for sampling.

Which sampling method should be selected for a particular BA analysis? It depends on the nature of the sample. As mentioned in the application notes in Table 5.4, the size of the population, the size of the sample, the area of application (geography, strata, ordering of the data, and so on), and even the researchers running the data collection effort impact the particular methodology selected. A best practices approach might begin with a determination of any constraints (time allowed and costs) that might limit the selection of a sample collection effort. That may narrow the choice to something like a quota method. Another best practices recommendation is to start with the objective(s) of the BA project and use them as a guide in the selection of the sampling method. For example, suppose the objective of a BA analysis is to increase sales of a particular product. This might lead to random sampling of customers or even a stratified sample by income levels, if income is important to the results of the analysis. Fortunately, there is software to make the data collection process easier and less expensive.

SAS software can be used with the methods mentioned earlier to aid in sampling analysis. For example, SAS permits simple, systematic, stratified, and cluster random methods, among others. Using this software requires a designation of the number of sample elements in each stratum. (For example, we selected 2 for each stratum in this example.) In Figure 5.11, SAS has defined seven strata for the Sales 4 data. The logic of this stratification can be observed by looking at the Sales 4 data in Figure 5.1, where only seven different types of values exist (1, 5, 9, 12, 18, 19, and 20). The additional SAS printout in Figure 5.11 shows the specific sample elements that were randomly selected in each stratum, as well as totals and their percentages in the resulting sample. For example, only 0.33, or 33 percent, of the "21" strata sample elements were randomly selected by the SAS program.

```
proc sort data= sales_data; by sales4;

proc surveyselect data =sales_data out = samp1 method = srs n=2;
strata sales4;
run;
```

```
                    The SAS System          17:53 Friday,

                 The SURVEYSELECT Procedure

        Selection Method      Simple Random Sampling
        Strata Variable       sales4

            Input Data Set           SALES_DATA
            Random Number Seed        522937001
            Stratum Sample Size               2
            Number of Strata                  7
            Total Sample Size                14
            Output Data Set              SAMP1
```

	sales4	Probability of Selection	Sampling Weight
1	1	1	1
2	1	1	1
3	5	1	1
4	5	1	1
5	9	1	1
6	9	1	1
7	12	1	1
8	12	1	1
9	18	1	1
10	18	1	1
11	19	0.5	2
12	19	0.5	2
13	21	0.3333333333	3
14	21	0.3333333333	3

Figure 5.11 SAS program statements for stratification/random sampling for Sales 4 variable

5.4.2 Sampling Estimation

Invariably, using any sampling method can cause errors in the sample results. Most of the statistical methods listed in Table 5.2 are formulated for population statistics. Once sampling is introduced into any statistical analysis, the data must be treated as a sample and not as a population. Many statistical techniques, such as standard error of mean and sample variance, incorporate mathematical correction factors to adjust descriptive analysis statistical tools to compensate for the possibility of sampling error.

One of the methods of compensating for error is to show some degree of confidence in any sampling statistic. The confidence in the sample statistics used can be expressed in a *confidence interval*, which is an interval estimate about the sample statistics. In general, we can express this interval estimate as follows:

Confidence interval = (sample statistic) ± [(confidence coefficient) × (standard error of the estimate)]

The *sample statistic* in the confidence interval can be any measure or proportion from a sample that is to be used to estimate a population parameter, such as a measure of central tendency like a mean. The *confidence coefficient* is set as a percentage to define the degree of confidence to accurately identify the correct sample statistic. The larger the confidence coefficient, the more likely the population mean from the sample will fall within the confidence interval. Many software systems set a 95 percent *confidence level* as the default confidence coefficient, although any percentage can be used. SAS permits the user to enter a desired percentage. The *standard error of the estimate* in the preceding expression can be any statistical estimate, including proportions used to estimate a population parameter. For example, using a mean as the sample statistic, we have the following interval estimate expression:

Confidence interval = mean ± [(95 percent) × (standard error of the mean)]

The output of this expression consists of two values that form high and low values defining the confidence interval. The interpretation of this interval is that the true population mean represented by the sample has a 95 percent chance of falling in the interval. In this way, there is still a 5 percent chance that the true population mean will not fall in the interval due to sampling error. Because the standard error of the mean is based on variation statistics (standard deviation), the larger the variance statistics used in this expression, the wider the confidence interval and the less precise the sample mean value, which results in a good estimate for the true population mean.

SAS computes confidence intervals when analyzing various statistical measures and tests. For example, the SAS summary printout in Table 5.5 is of the 95 percent confidence interval for the Sales 1 variable. With a sample mean value of 35.15, the confidence interval suggests there is a 95 percent chance that the true population mean falls between 29.91 and 40.39. When trying to ascertain if the sample is of any value, this kind of information can be of great significance. For example, knowing with 95 percent certainty there is at least a mean of 29.91 might make the difference between continuing to sell a product or not because of a needed requirement for a breakeven point in sales.

Table 5.5 SAS 95 Percent Confidence Interval Summary for Sales 1 Variable

One-Sample Statistics						
	N	Mean	Std. Deviation	Std. Error Mean	95 Percent Confidence Interval of the Difference	
Sales 1	20	35.15	11.198	2.504	Lower	Upper
					29.91	40.39

Confidence intervals are also important for demonstrating the accuracy of some forecasting models. For example, confidence intervals can be created about a regression equation model forecast to see how far off the estimates might be if the model is used to predict future sales. For additional discussion on confidence intervals, see Appendix A, "Statistical Tools."

5.5 Introduction to Probability Distributions

By taking samples, one seeks to reveal population information. Once a sample is taken on which to base a forecast or a decision, it may not accurately capture the population information. No single sample can assure an analyst that the true population information has been captured. Confidence interval statistics are employed to reflect the possibility of error from the true population information.

To utilize the confidence interval formula expressed in Section 5.4, you set a confidence coefficient percentage (95 percent) as a way to express the possibility that the sample statistics used to represent the population statistics may have a potential for error. The confidence coefficient used in the confidence interval is usually referred to as a Z *value*. It is spatially related to the area (expressed as a percentage or frequency) representing the probability under the curve of a distribution. The *sample standard normal distribution* is the bell-shaped curve illustrated in Figure 5.12. This distribution shows the relationship of the Z value to the area under the curve. The Z value is the number of standard error of the means.

The *confidence coefficient* is related to the Z values, which divide the area under a normal curve into probabilities. Based on the *central limit theorem*, we assume that all sampling distributions of sufficient size are normally distributed with a standard deviation equal to the standard error of the estimate. This means that an interval of plus or minus two standard errors of the estimate (whatever the estimate is) has a 95.44 percent chance of containing the true or actual population parameter. Plus or minus three standard errors of the estimate has a 99.74 percent chance of containing the true or actual population parameter. So the Z value represents the number of standard errors of the estimate. Table 5.6 has selected Z values for specific confidence levels representing the probability that the true population parameter is within the confidence interval and represents the percentage of area under the curve in that interval.

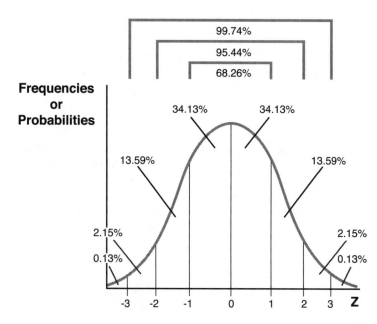

Figure 5.12 Standard normal probability distribution

Table 5.6 Selected Z Values and Confidence Levels

Confidence Level	Related Z Value
0.60	0.253
0.70	0.524
0.80	0.842
0.90	1.282
0.95	1.645
0.99	2.327
0.999	3.080

The important BA use of the probability distributions and confidence intervals is that they suggest an assumed parameter based on a sample that has properties that allow analysts to predict or forecast with some assessed degree of statistical accuracy. In other words, BA analysts can, with some designated confidence level, use samples from large databases to accurately predict population parameters.

Another important value to probability distributions is that they can be used to compute probabilities that certain outcomes like success with business performance may occur. In the exploratory descriptive analytics step of the BA process, assessing the probabilities of some events occurring can be a useful strategy to guide subsequent steps in an analysis. Indeed, probability information may be useful in weighing the choices an analyst faces in any of the steps of the BA process. Suppose, for example, the statistics from the Sales 1 variable in Table 5.5 are treated as a sample to discover the probability of sales greater than one standard error of the mean above the current mean of 35.15. In Figure 5.13, the mean (35.15) and standard error of the mean (2.504) statistics are included at the bottom of the standard sampling normal distribution. When one standard error of the mean is added to the sample mean, the resulting value is 37.654. The sum of the area (the shaded region in Figure 5.13) representing the total probability beyond 37.654 is a probability of 15.87 (13.59+2.15+0.13). So there is only a 15.87 percent probability that sales will exceed 37.654 based on the sample information for the Sales 1 variable.

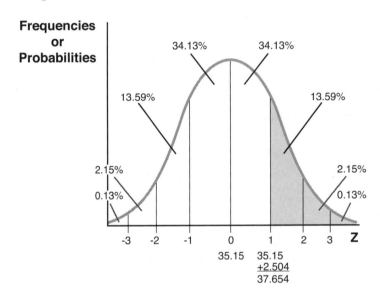

Figure 5.13 Probability function example

The ability to assess probabilities using this approach is applicable to other types of probability distributions. For a review of probability concepts and distributions, probability terminology, and probability applications, see Appendix A.

5.6 Marketing/Planning Case Study Example: Descriptive Analytics Step in the BA Process

In the last section of this chapter and in Chapters 6, "What Is Predictive Analytics?" and 7, "What Is Prescriptive Analytics?" an ongoing marketing/planning case study of the relevant BA step discussed in those chapters will be presented to illustrate some of the tools and strategies used in a BA problem analysis. This is the first installment of the case study dealing with the descriptive analytics step in BA. The predictive analytics step (in Chapter 6) and prescriptive analytics step (in Chapter 7) will continue with this ongoing case study.

5.6.1 Case Study Background

A firm has collected a random sample of monthly sales information on a service product offered infrequently and only for a month at a time. The sale of this service product occurs only during the month that the promotion efforts are allocated. Basically, promotion funds are allocated at the beginning or during the month, and whatever sales occur are recorded for that promotion effort. There is no spillover of promotion to another month, because monthly offerings of the service product are independent and happen randomly during any particular year. The nature of the product does not appear to be impacted by seasonal or cyclical variations, which prevents forecasting and makes planning the budget difficult.

The firm promotes this service product by using radio commercials, newspaper ads, television commercials, and point-of-sale (POS) ad cards. The firm has collected the sales information as well as promotion expenses. Because the promotion expenses are put into place before the sales take place and on the assumption that the promotion efforts impact products, the four promotion expenses can be viewed as predictive data sets (or what will be the predictive variables in a forecasting model). Actually, in terms of modeling this problem, product sales is going to be considered the dependent variable, and the other four data sets represent independent or predictive variables.

These five data sets, in thousands of dollars, are present in the SAS printout shown in Figure 5.14. What the firm would like to know is, given a fixed budget of $350,000 for promoting this service product, when offered again, how best should budget dollars be allocated in the hope of maximizing future estimated months' product sales? This is a typical question asked of any product manager and marketing manager's promotion efforts. Before the firm allocates the budget, there is a need to understand how to estimate future product sales. This requires understanding the behavior of product sales relative to sales promotion. To begin to learn about the behavior of product sales to promotion efforts, we begin with the first step in the BA process: descriptive analytics.

case_number	sales	radio	paper	tv	pos
1	1	11125	65	89	250	1.3
2	2	16121	73	55	260	1.6
3	3	16440	74	58	270	1.7
4	4	16876	75	82	270	1.3
5	5	13965	69	75	255	1.5
6	6	14999	70	71	255	2.1
7	7	20167	87	59	280	1.2
8	8	20450	89	65	280	3
9	9	15789	72	62	260	1.6
10	10	15991	73	56	260	1.6
11	11	15234	70	66	255	1.5
12	12	17522	78	50	270	0
13	13	17933	79	47	275	0.2
14	14	18390	81	78	275	0.9
15	15	18723	81	41	275	1
16	16	19328	84	63	280	2.6
17	17	19399	84	77	280	1.2
18	18	19641	85	35	280	2.5
19	19	12369	65	37	250	2.5
20	20	13882	68	80	252	1.4

Figure 5.14 Data for marketing/planning case study

5.6.2 Descriptive Analytics Analysis

To begin conceptualizing possible relationships in the data, one might compute some descriptive statistics and graph charts of data (which will end up being some of the variables in the planned model). SAS can be used to compute these statistics and charts. The SAS printout in Table 5.7 provides a typical set of basic descriptive statistics (means, ranges, standard deviations, and so on) and several charts.

				The MEANS Procedure			
Variable	N	Range	Minimum	Maximum	Mean	Std Dev	Variance
radio	20	24.0000000	65.0000000	89.0000000	76.1000000	7.3549124	54.0947368
paper	20	54.0000000	35.0000000	89.0000000	62.3000000	15.3592078	235.9052632
tv	20	30.0000000	250.0000000	280.0000000	266.6000000	11.3388016	128.5684211
pos	20	3.0000000	0	3.0000000	1.5350000	0.7499298	0.5623947
sales	20	9325.00	11125.00	20450.00	16717.20	2617.05	6848960.59

Table 5.7 SAS Descriptive Statistics for the Marketing/Planning Case Study

Remember, this is the beginning of an exploration that seeks to describe the data and get a handle on what it may reveal. This effort may take some exploration to figure out the best way to express data from a file or database, particularly as the size of the data file increases. In this simple example, the data sets are small but can still reveal valuable information if explored well.

In Figure 5.15, five typical SAS charts are presented. Respectively, these charts include a histogram chart (sales), a block chart (radio), a line chart (TV), a pie chart (paper), and a 3D chart (POS). These charts are interesting, but they're not very revealing of behavior that helps in understanding future sales trends that may be hiding in this data.

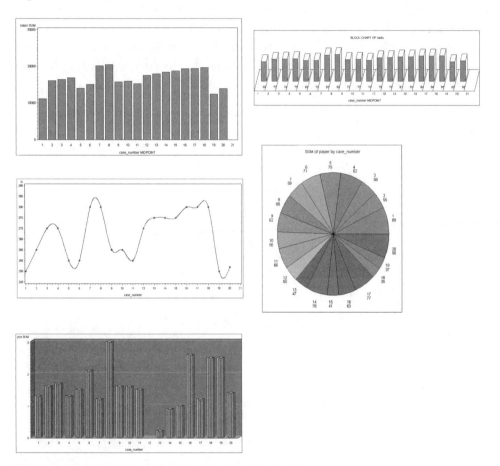

Figure 5.15 Preliminary SAS charts for the marketing/planning case study

To expedite the process of revealing potential relational information, think in terms of what one is specifically seeking. In this instance, it is to predict the future sales of the service product. That means looking for a graph to show a trend line. One type of simple graph that is related to trend analysis is a line chart. Using SAS again, one can compute line charts for each of the five data sets. These charts are presented in Figure 5.16. The vertical axis consists of the dollar values, and the horizontal axis is the number ordering of observations as listed in the data sets.

While providing a less confusing graphic presentation of the up-and-down behavior of the data, the charts in these figures still do not clearly reveal any possible trend information. Because the 20 months of data are not in any particular order and are not related to time, they are independent values that can be reordered. Reordering data or sorting it can be a part of the descriptive analytics process. Because trend is usually an upward or downward linear behavior, one might be able to observe a trend in the product sales data set if that data is reordered from low to high (or high to low). Reordering the sales by moving the 20 rows of data around such that sales is arranged from low to high is presented in Figure 5.17. Using this reordered data set, the SAS results are illustrated in the new line charts in Figure 5.18.

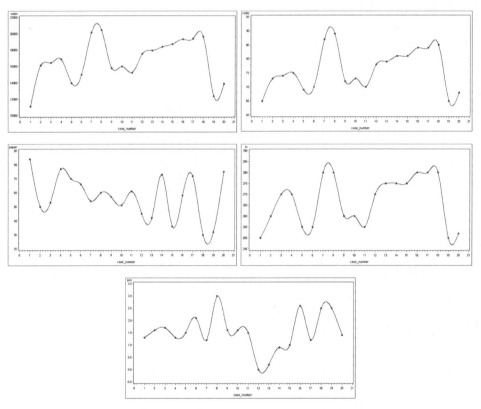

Figure 5.16 Preliminary SAS line charts for the marketing/planning case study

	case_number	sales	radio	paper	tv	pos
1	1	11125	65	89	250	1.3
2	19	12369	65	37	250	2.5
3	20	13882	68	80	252	1.4
4	5	13965	69	75	255	1.5
5	6	14999	70	71	255	2.1
6	11	15234	70	66	255	1.5
7	9	15789	72	62	260	1.6
8	10	15991	73	56	260	1.6
9	2	16121	73	55	260	1.6
10	3	16440	74	58	270	1.7
11	4	16876	75	82	270	1.3
12	12	17522	78	50	270	0
13	13	17933	79	47	275	0.2
14	14	18390	81	78	275	0.9
15	15	18723	81	41	275	1
16	16	19328	84	63	280	2.6
17	17	19399	84	77	280	1.2
18	18	19641	85	35	280	2.5
19	7	20167	87	59	280	1.2
20	8	20450	89	65	280	3

Figure 5.17 Reordered data in line charts for the marketing/planning case study

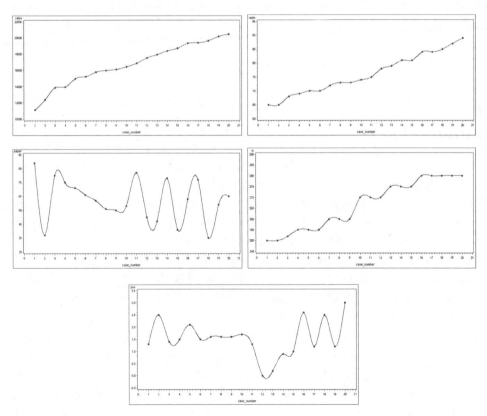

Figure 5.18 SAS line charts based on reordered data for the marketing/planning case study

Given the low to high reordering of the product sales as a guide, some of the other four line charts suggest a relationship with product sales. Both radio and TV commercials appear to have a similar low to high trending relationship that matches product sales. This suggests these two will be good predictive variables for product sales, whereas newspaper and POS ads are still volatile in their charted relationships with product sales. Therefore, these two latter variables might not be useful in a model seeking to predict product sales. They cannot be ruled out at this point in the analysis, but they are suspected of adding little to a model for accurately forecasting product sales. Put another way, they appear to add unneeded variation that may take away from the accuracy of the model. Further analysis is called for to explore in more detail and sophistication the best set of predictive variables to predict the relationships in product sales.

In summary, for this case study, the descriptive analytics analysis has revealed a potential relationship between radio and TV commercials and future product sales, and it questions the relationship of newspaper and POS ads to sales. The managerial ramifications of these results might suggest discontinuing investing in newspaper and POS ads and more productively allocating funds to radio and TV commercials. Before such a reallocation can be justified, more analysis is needed. The next step in the analysis, predictive analytics, is presented in the last section of Chapter 6.

Summary

This chapter discussed data visualization and exploration. In particular, this chapter described and illustrated graphic and statistical methods useful in the descriptive analytics step of the BA process. Illustrations of SAS printouts of graphs, charts, and statistical methods were presented. In addition, sampling methods were described, along with the available software applications from SAS. Sampling estimation was also discussed, as was its connection to sampling distributions for purposes of error estimation in measures of central tendency. Finally, this chapter presented the first installment of a case study illustrating the descriptive analytics step of the BA process. The remaining installments are presented in Chapters 6 and 7.

Several of the appendixes of this book are designed to augment the chapter material by including technical, mathematical, and statistical tools. For both greater understanding of the methodologies discussed in this chapter and a basic review of statistical and other quantitative methods, a review of the appendixes mentioned in this chapter is recommended.

The results of the descriptive analytics step of the BA process create an exploratory foundation on which further analysis can be based. In Chapter 6, we continue with the second step of the BA process: predictive analytics.

Discussion Questions

1. Why is it important to explore data with graphs and charts?
2. What is the difference between skewedness and kurtosis?
3. Why would we ever want to use a sample if we have population information?
4. Is there a way to determine skewedness from the ordering of the mean, median, and mode measures of central tendency?
5. Which of the sampling methods listed in Table 5.4 is the best, and why?
6. In setting the confidence level, why not just set one that is low enough for the population parameter to be assured of inclusion?

Problems

1. Using SAS, draw a line graph of the Sales 2 distribution from the data in Figure 5.1 and compute the mean, median, and mode statistics. Does the positioning of the mean, median, and mode support the skewedness statistic from your computations of the mean, median, and mode? Explain the answer.
2. Using SAS, draw a scatter diagram of the Sales 3 distribution from the data in Figure 5.1 and compute the mean, median, and mode statistics. Does the positioning of the mean, median, and mode support the skewedness statistic from your computations of the mean, median, and mode? Explain the answer.
3. Using SAS, make a random sample on Sales 3 distribution from the data in Figure 5.1. Using the software, determine four items from the data set for sampling purposes. Which specific values should be selected from the data set?
4. Using SAS, make a random sample on Sales 2 distribution from the data in Figure 5.1. Using the software, determine six items from the data set for sampling purposes. Which specific values should be selected from the data set?
5. With a mean value of 50 and a standard error of the mean of 12, what is the 90 percent confidence interval for this problem?

6. With a mean value of 120 and a standard error of the mean of 20, what is the 99 percent confidence interval for this problem?

7. A firm has computed its mean sales for a new product to be 2,000 units for the year, with a standard error of the mean of 56. The firm would like to know if the probability of its mean sales for next year (based on this year) will be above 2,112. What is the probability?

8. The Homes Golf Ball Company has made a number of different golf products over the years. Research on thousands of balls revealed the mean flight distance of its Maximum Fly golf ball product to be 450 yards, with a standard error of the mean of 145 yards. The company is hoping to improve the product to fly an additional 290 yards. What is the probability of the improvement from 450 to 740 yards?

6

What Is Predictive Analytics?

Chapter objectives:

- Explain what logic-driven models are used for in business analytics (BA).
- Describe what a cause-and-effect diagram is used for in BA.
- Explain the difference between logic-driven and data-driven models.
- Explain how data mining can aid in BA.
- Explain why neural networks can be helpful in determining both associations and classification tasks required in some BA analyses.
- Explain how clustering is undertaken in BA.
- Explain how step-wise regression can be useful in BA.
- Explain how to use R-Squared adjusted statistics in BA.

6.1 Introduction

In Chapter 1, "What Is Business Analytics?" we defined predictive analytics as an application of advanced statistical, information software, or operations research methods to identify predictive variables and build predictive models to identify trends and relationships not readily observed in the descriptive analytic analysis. Knowing that relationships exist explains why one set of independent variables (predictive variables) influences dependent variables like business performance. Chapter 1 further explained that the purpose of the descriptive analytics step is to position decision makers to build predictive models designed to identify and predict future trends.

Picture a situation in which big data files are available from a firm's sales and customer information (responses to differing types of advertisements, customer surveys on product quality, customer surveys on supply chain performance, sale prices, and so on). Assume also that a previous descriptive analytic analysis suggests that there is a relationship between certain customer variables, but there is a need to precisely establish a quantitative relationship between sales and customer behavior. Satisfying this need requires exploration into the big data to first establish whether a measurable, quantitative relationship does in fact exist and then develop a statistically valid model in which to predict future events. This is what the predictive analytics step in BA seeks to achieve.

Many methods can be used in this step of the BA process. Some are just to sort or classify big data into manageable files in which to later build a precise quantitative model. As previously mentioned in Chapter 3, "What Resource Considerations Are Important to Support Business Analytics?" predictive modeling and analysis might consist of the use of methodologies, including those found in forecasting, sampling and estimation, statistical inference, data mining, and regression analysis. A commonly used methodology is multiple regression. (See Appendixes A, "Statistical Tools," and E, "Forecasting," for a discussion on multiple regression and ANOVA testing.) This methodology is ideal for establishing whether a statistical relationship exists between the predictive variables found in the descriptive analysis and the dependent variable one seeks to forecast. An example of its use will be presented in the last section of this chapter.

Although single or multiple regression models can often be used to forecast a trend line into the future, sometimes regression is not practical. In such cases, other forecasting methods, such as exponential smoothing or smoothing averages, can be applied as predictive analytics to develop needed forecasts of business activity. (See Appendix E.) Whatever methodology is used, the identification of future trends or forecasts is the principal output of the predictive analytics step in the BA process.

6.2 Predictive Modeling

Predictive modeling means developing models that can be used to forecast or predict future events. In business analytics, models can be developed based on logic or data.

6.2.1 Logic-Driven Models

A *logic-driven model* is one based on experience, knowledge, and logical relationships of variables and constants connected to the desired business performance outcome situation. The question here is how to put variables and constants together to create a model that can predict the future. Doing this requires business experience.

Model building requires an understanding of business systems and the relationships of variables and constants that seek to generate a desirable business performance outcome. To help conceptualize the relationships inherent in a business system, diagramming methods can be helpful. For example, the *cause-and-effect diagram* is a visual aid diagram that permits a user to hypothesize relationships between potential causes of an outcome (see Figure 6.1). This diagram lists potential causes in terms of human, technology, policy, and process resources in an effort to establish some basic relationships that impact business performance. The diagram is used by tracing contributing and relational factors from the desired business performance goal back to possible causes, thus allowing the user to better picture sources of potential causes that could affect the performance. This diagram is sometimes referred to as a *fishbone diagram* because of its appearance.

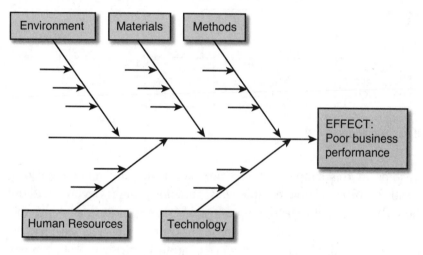

Figure 6.1   Cause-and-effect diagram*

Source: Adapted from Figure 5 in Schniederjans et al. (2014), p. 201.

Another useful diagram to conceptualize potential relationships with business performance variables is called the *influence diagram*. According to Evans (2013, pp. 228–229), influence diagrams can be useful to conceptualize the relationships of variables in the development of models. An example of an influence diagram is presented in Figure 6.2. It maps the relationship of variables and a constant to the desired business performance outcome of profit. From such a diagram, it is easy to convert the information into a quantitative model with constants and variables that define profit in this situation:

Profit = Revenue – Cost, or

Profit = (Unit Price × Quantity Sold) – [(Fixed Cost) +
(Variable Cost × Quantity Sold)], or

P = (UP × QS) – [FC + (VC × QS)]

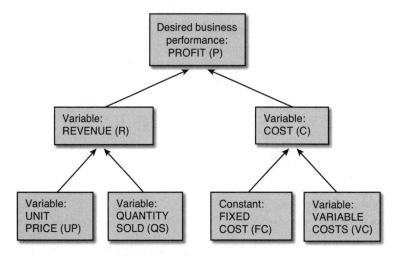

Figure 6.2 An influence diagram

The relationships in this simple example are based on fundamental business knowledge. Consider, however, how complex cost functions might become without some idea of how they are mapped together. It is necessary to be knowledgeable about the business systems being modeled in order to capture the relevant business behavior. Cause-and-effect diagrams and influence diagrams provide tools to conceptualize relationships, variables, and constants, but it often takes many other methodologies to explore and develop predictive models.

6.2.2 Data-Driven Models

Logic-driven modeling is often used as a first step to establish relationships through *data-driven models* (using data collected from many sources to quantitatively establish model relationships). To avoid duplication of content and focus on conceptual material in the chapters, we have relegated most of the computational aspects and some computer usage content to the appendixes. In addition, some of the methodologies are illustrated in the case problems presented in this book. Please refer to the Additional Information column in Table 6.1 to obtain further information on the use and application of the data-driven models.

Table 6.1 Data-Driven Models

Data-Driven Models	Possible Applications	Additional Information
Sampling and Estimation	Generate statistical confidence intervals to define limitations and boundaries on future forecasts for other forecasting models.	Chapter 5, "What Is Descriptive Analytics?" Appendix A, Appendix E.
Regression Analysis	(1) Create a predictive equation useful for forecasting time series forecasts. (2) Weed out predictive variables in forecasting models that add little to predicting values. (3) Generate a trend line for forecasting.	Chapter 6, "What Is Predictive Analytics?" Chapter 8, "A Final Business Analytics Case Problem," Appendix E.
Correlation Analysis	(1) Assess variable relationships. (2) Weed out predictive variables in forecasting models that add little to predicting values.	Chapter 6, Appendix E.
Probability Distributions	(1) Estimate trend behavior that follows certain types of probability distributions. (2) Conduct statistical tests to confirm significance of variables.	Chapter 5, Appendix A.
Predictive Modeling and Analysis	Fit linear and nonlinear models to data to use the models for forecasting.	Appendix A, Appendix E.
Forecasting Models	Those listed in this table and others such as smoothing models can be used to forecast values.	Appendix E.
Simulation	Project future behavior in variables by simulating the past behavior found in probability distributions.	Appendix F, "Simulation."

6.3 Data Mining

As mentioned in Chapter 3, *data mining* is a discovery-driven software application process that provides insights into business data by finding hidden patterns and relationships in big or small data and inferring rules from them to predict future behavior.

These observed patterns and rules guide decision-making. This is not just numbers, but text and social media information from the Web. For example, Abrahams et al. (2013) developed a set of text-mining rules that automobile manufacturers could use to distill or mine specific vehicle component issues that emerge on the Web but take months to show up in complaints or other damaging media. These rules cut through the mountainous data that exists on the Web and are reported to provide marketing and competitive intelligence to manufacturers, distributors, service centers, and suppliers. Identifying a product's defects and quickly recalling or correcting the problem before customers experience a failure reduce customer dissatisfaction when problems occur.

6.3.1 A Simple Illustration of Data Mining

Suppose a grocery store has collected a big data file on what customers put into their baskets at the market (the collection of grocery items a customer purchases at one time). The grocery store would like to know if there are any associated items in a typical market basket. (For example, if a customer purchases product A, she will most often associate it or purchase it with product B.) If the customer generally purchases product A and B together, the store might only need to advertise product A to gain both product A's and B's sales. The value of knowing this association of products can improve the performance of the store by reducing the need to spend money on advertising both products. The benefit is real if the association holds true.

Finding the association and proving it to be valid require some analysis. From the descriptive analytics analysis, some possible associations may have been uncovered, such as product A's and B's association. With any size data file, the normal procedure in data mining would be to divide the file into two parts. One is referred to as a training data set, and the other as a validation data set. The *training data set* develops the association rules, and the *validation data set* tests and proves that the rules work. Starting with the training data set, a common data mining methodology is *what-if analysis* using logic-based software. SAS has a what-if logic-based software application, and so do a number of other software vendors (see Chapter 3). These software applications allow logic expressions. (For example, if product A is present, then is product B present?) The systems can also provide frequency and probability information to show the strength of the association. These software systems have differing capabilities, which permit users to deterministically simulate different scenarios to identify complex combinations of associations between product purchases in a market basket.

Once a collection of possible associations is identified and their probabilities are computed, the same logic associations (now considered association rules) are rerun using the validation data set. A new set of probabilities can be computed, and those can be statistically compared using hypothesis testing methods to determine their similarity. Other software systems compute correlations for testing purposes to judge the strength and the direction of the relationship. In other words, if the consumer buys product A first, it could be referred to as the *Head* and product B as the *Body* of the association (Nisbet et al., 2009, p. 128). If the same basic probabilities are statistically significant, it lends validity to the association rules and their use for predicting market basket item purchases based on groupings of products.

6.3.2 Data Mining Methodologies

Data mining is an ideal predictive analytics tool used in the BA process. We mentioned in Chapter 3 different types of information that data mining can glean, and Table 6.2 lists a small sampling of data mining methodologies to acquire different types of information. Some of the same tools used in the descriptive analytics step are used in the predictive step but are employed to establish a model (either based on logical connections or quantitative formulas) that may be useful in predicting the future.

Table 6.2 Types of Information and Data Mining Methodologies

Types of Information	Description	Sample of Data Mining Methodologies
Association	Occurrence linked to a single event.	Association rules (for example, if-then analysis), correlation analysis, neural networks.
Classification	Pattern that describes the group an item belongs to. Found by examining previous classified existing items and inferring a set of rules that guide the classification process.	Discriminant analysis, logistics regression, neural networks.
Clustering	Similar to classification when no groups have yet been defined. Helps discover different groupings within data.	Hierarchical clustering, K-mean clustering.
Forecasting	Used to predict values that can identify patterns in customer behavior.	Regression analysis, correlation analysis.
Sequence	Event that is linked over time.	Lag correlation analysis, cause-and-effect diagrams.

Several computer-based methodologies listed in Table 6.2 are briefly introduced here. *Neural networks* are used to find associations where connections between words or numbers can be determined. Specifically, neural networks can take large volumes of data and potential variables and explore variable associations to express a beginning variable (referred to as an *input layer*), through middle layers of interacting variables, and finally to an ending variable (referred to as an *output*). More than just identifying simple one-on-one associations, neural networks link multiple association pathways through big data like a collection of nodes in a network. These nodal relationships constitute a form of classifying groupings of variables as related to one another, but even more, related in complex paths with multiple associations (Nisbet et al., 2009, pp. 128–138). Differing software have a variety of association network function capabilities. SAS offers a series of search engines that can identify associations. SPSS has two versions of neural network software functions: *Multilayer Perception* (MLP) and *Radial Basis Function* (RBF). Both procedures produce a predictive model for one or more dependent variables based on the values of the predictive variables. Both allow a decision maker to develop, train, and use the software to identify particular traits (such as bad loan risks for a bank) based on characteristics from data collected on past customers.

Discriminant analysis is similar to a multiple regression model except that it permits continuous independent variables and a categorical dependent variable. The analysis generates a regression function whereby values of the independent variables can be incorporated to generate a predicted value for the dependent variable. Similarly, *logistic regression* is like multiple regression. Like discriminant analysis, its dependent variable can be categorical. The independent variables in logistic regression can be either continuous or categorical. For example, in predicting potential outsource providers, a firm might use a logistic regression, in which the dependent variable would be to classify an outsource provider as either rejected (represented by the value of the dependent variable being zero) or acceptable (represented by the value of one for the dependent variable).

Hierarchical clustering is a methodology that establishes a hierarchy of clusters that can be grouped by the hierarchy. Two strategies are suggested for this methodology: agglomerative and divisive. The *agglomerative strategy* is a bottom-up approach, in which one starts with each item in the data and begins to group them. The *divisive strategy* is a top-down approach, in which one starts with all the items in one group and divides the group into clusters. How the clustering takes place can involve many different types of algorithms and differing software applications. One method commonly used is to employ a Euclidean distance formula that looks at the square root of the sum of distances between two variables, their differences squared. Basically, the formula seeks to match up variable candidates that have the least squared error differences. (In other words, they're closer together.)

K-mean clustering is a classification methodology that permits a set of data to be reclassified into *K groups*, where *K* can be set as the number of groups desired. The algorithmic process identifies initial candidates for the *K* groups and then interactively searches other candidates in the data set to be averaged into a mean value that represents a particular *K* group. The process of selection is based on maximizing the distance from the initial *K* candidates selected in the initial run through the list. Each run or iteration through the data set allows the software to select further candidates for each group.

The K-mean clustering process provides a quick way to classify data into differentiated groups. To illustrate this process, use the sales data in Figure 6.3 and assume these are sales from individual customers. Suppose a company wants to classify the sales customers into high and low sales groups.

	time	sale
1	1	13444
2	2	12369
3	3	15322
4	4	13965
5	5	14999
6	6	15234
7	7	12999
8	8	15991
9	9	16121
10	10	18654
11	11	16876
12	12	17522
13	13	17933
14	14	15233
15	15	18723
16	16	13855
17	17	19399
18	18	16854
19	19	20167
20	20	18654

Figure 6.3 Sales data for cluster classification problem

The SAS K-Mean cluster software can be found in Proc Cluster. Any integer value can designate the *K* number of clusters desired. In this problem set, *K*=2. The SAS printout of this classification process is shown in Table 6.3. The Initial Cluster Centers table listed the initial high (20167) and a low (12369) value from the data set as the clustering process begins. As it turns out, the software divided the customers into 9 high sales customers and 11 low sales customers.

```
                        The FASTCLUS Procedure
            Replace=FULL  Radius=0  Maxclusters=2 Maxiter=1

                            Initial Seeds

                Cluster             time             sale

                   1            19.00000        20167.00000
                   2             2.00000        12369.00000


            Criterion Based on Final Seeds  =    797.6


                          Cluster Summary

                              Maximum Distance
                       RMS Std       from Seed    Radius    Nearest   Distance Between
Cluster   Frequency  Deviation   to Observation  Exceeded   Cluster   Cluster Centroids
   1           9        789.8          1857.9                  2            3806.2
   2          11        879.3          2133.9                  1            3806.2
```

Table 6.3 SAS K-Mean Cluster Solution

Consider how large big data sets can be. Then realize this kind of classification capability can be a useful tool for identifying and predicting sales based on the mean values.

There are so many BA methodologies that no single section, chapter, or even book can explain or contain them all. The analytic treatment and computer usage in this chapter have been focused mainly on conceptual use. For a more applied use of some of these methodologies, note the case study that follows and some of the content in the appendixes.

6.4 Continuation of Marketing/Planning Case Study Example: Prescriptive Analytics Step in the BA Process

In the last sections of Chapters 5, 6, and 7, an ongoing marketing/planning case study of the relevant BA step discussed in those chapters is presented to illustrate some of the tools and strategies used in a BA problem analysis. This is the second installment of the case study dealing with the predictive analytics analysis step in BA. The prescriptive analysis step coming in Chapter 7, "What Is Prescriptive Analytics?" will complete the ongoing case study.

6.4.1 Case Study Background Review

The case study firm had collected a random sample of monthly sales information presented in Figure 6.4 listed in thousands of dollars. What the firm wants to know is, given a fixed budget of $350,000 for promoting this service product, when it is offered again, how best should the company allocate budget dollars in hopes of maximizing the future estimated month's product sales? Before the firm makes any allocation of budget, there is a need to understand how to estimate future product sales. This requires understanding the behavior of product sales relative to sales promotion efforts using radio, paper, TV, and point-of-sale (POS) ads.

case_number	sales	radio	paper	tv	pos
1	1	11125	65	89	250	1.3
2	2	16121	73	55	260	1.6
3	3	16440	74	58	270	1.7
4	4	16876	75	82	270	1.3
5	5	13965	69	75	255	1.5
6	6	14999	70	71	255	2.1
7	7	20167	87	59	280	1.2
8	8	20450	89	65	280	3
9	9	15789	72	62	260	1.6
10	10	15991	73	56	260	1.6
11	11	15234	70	66	255	1.5
12	12	17522	78	50	270	0
13	13	17933	79	47	275	0.2
14	14	18390	81	78	275	0.9
15	15	18723	81	41	275	1
16	16	19328	84	63	280	2.6
17	17	19399	84	77	280	1.2
18	18	19641	85	35	280	2.5
19	19	12369	65	37	250	2.5
20	20	13882	68	80	252	1.4

Figure 6.4 Data for marketing/planning case study

The previous descriptive analytics analysis in Chapter 5 revealed a potentially strong relationship between radio and TV commercials that might be useful in predicting future product sales. The analysis also revealed little regarding the relationship of newspaper and POS ads to product sales. So although radio and TV commercials are most promising, a more in-depth predictive analytics analysis is called for to accurately measure and document the degree of relationship that may exist in the variables to determine the best predictors of product sales.

6.4.2 Predictive Analytics Analysis

An ideal multiple variable modeling approach that can be used in this situation to explore variable importance in this case study and eventually lead to the development of a predictive model for product sales is correlation and multiple regression. We will use SAS's statistical package to compute the statistics in this step of the BA process.

First, we must consider the four independent variables—radio, TV, newspaper, POS—before developing the model. One way to see the statistical direction of the relationship (which is better than just comparing graphic charts) is to compute the

Pearson correlation coefficients r between each of the independent variables with the dependent variable (product sales). The SAS correlation coefficients and their levels of significance are presented in Table 6.4. The larger the Pearson correlation (regardless of the sign) and the smaller the *Significance test* values (these are t-tests measuring the significance of the Pearson r value; see Appendix A), the more significant the relationship. Both radio and TV are statistically significant correlations, whereas at a 0.05 level of significance, paper and POS are not statistically significant.

Table 6.4 SAS Pearson Correlation Coefficients: Marketing/Planning Case Study

Statistic	Radio	Paper	TV	POS
Pearson Correlation r with Product Sales	.977	−.283	.958	.013
Significance Test (1-Tailed)°	.000	.113	.000	.479

```
                   Pearson Correlation Coefficients, N = 20
                         Prob > |r| under H0: Rho=0

                sales        radio        paper          tv          pos

   sales      1.00000      0.97714     -0.28307      0.95797      0.01265
                           <.0001        0.2265       <.0001       0.9578

   radio      0.97714      1.00000     -0.23836      0.96610      0.06040
              <.0001                     0.3115       <.0001       0.8003

   paper     -0.28307     -0.23836      1.00000     -0.24588     -0.09006
               0.2265       0.3115                    0.2960       0.7057

   tv         0.95797      0.96610     -0.24588      1.00000     -0.03602
              <.0001        <.0001       0.2960                    0.8802

   pos        0.01265      0.06040     -0.09006     -0.03602      1.00000
               0.9578       0.8003       0.7057       0.8802
```

*Values of 0.05 or less would designate a significant relationship with product sales

Although it can be argued that the positive or negative correlation coefficients should not automatically discount any variable from what will be a predictive model, the negative correlation of newspapers suggests that as a firm increases investment in newspaper ads, it will decrease product sales. This does not make sense in this case study. Given the illogic of such a relationship, its potential use as an independent variable in a model is questionable. Also, this negative correlation poses several questions that should be considered. Was the data set correctly collected? Is the data set accurate? Was the sample large enough to have included enough data for this variable to show a positive relationship? Should it be included for further analysis? Although it is possible that a negative relationship can statistically show up like this, it does not make sense in this case. Based on this reasoning and the fact that the correlation is not statistically significant, this variable (newspaper ads) will be removed from further consideration in this exploratory analysis to develop a predictive model.

Some researchers might also exclude POS based on the insignificance ($p=0.479$) of its relationship with product sales. However, for purposes of illustration, continue to consider it a candidate for model inclusion. Also, the other two independent variables (radio and TV) were found to be significantly related to product sales, as reflected in the correlation coefficients in the tables.

At this point, there is a dependent variable (product sales) and three candidate independent variables (POS, TV, and Radio) in which to establish a predictive model that can show the relationship between product sales and those independent variables. Just as a line chart was employed to reveal the behavior of product sales and the other variables in the descriptive analytic step, a statistical method can establish a linear model that combines the three predictive variables. We will use multiple regression, which can incorporate any of the multiple independent variables, to establish a relational model for product sales in this case study. Multiple regression also can be used to continue our exploration of the candidacy of the three independent variables.

The procedure by which multiple regression can be used to evaluate which independent variables are best to include or exclude in a linear model is called *step-wise multiple regression*. It is based on an evaluation of regression models and their validation statistics—specifically, the multiple correlation coefficients and the F-ratio from an ANOVA. SAS software and many other statistical systems build in the step-wise process. Some are called *backward selection* or *step-wise regression*, and some are called *forward selection* or *step-wise regression*. The backward step-wise regression starts with all the independent variables placed in the model, and the step-wise process removes them one at a time based on worst predictors first until a statistically significant model emerges. The forward step-wise regression starts with the best related variable (using correction analysis as a guide), and then step-wise adds other variables until adding more will no longer improve the accuracy of the model. The forward step-wise regression process will be illustrated here manually. The first step is to generate individual regression models and statistics for each independent variable with the dependent variable one at a time. These three SAS models are presented in Tables 6.5, 6.6, and 6.7 for the POS, radio, and TV variables, respectively.

```
                        The REG Procedure
                         Model: MODEL1
                     Dependent Variable: sales

              Number of Observations Read        20
              Number of Observations Used        20

                       Analysis of Variance

                                Sum of          Mean
    Source              DF      Squares         Square     F Value   Pr > F

    Model                1        20819          20819        0.00   0.9578
    Error               18    130109432        7228302
    Corrected Total     19    130130251


            Root MSE            2688.55013    R-Square      0.0002
            Dependent Mean           16717    Adj R-Sq     -0.0554
            Coeff Var           16.08254


                       Parameter Estimates

                         Parameter      Standard
    Variable      DF      Estimate        Error     t Value   Pr > |t|

    Intercept      1         16649    1398.32203       11.91    <.0001
    pos            1      44.14019     822.47123        0.05    0.9578
```

Table 6.5 SAS POS Regression Model: Marketing/Planning Case Study

```
                        The REG Procedure
                         Model: MODEL1
                     Dependent Variable: sales

              Number of Observations Read        20
              Number of Observations Used        20

                       Analysis of Variance

                                Sum of          Mean
    Source              DF      Squares         Square     F Value   Pr > F

    Model                1    124248210      124248210      380.22    <.0001
    Error               18      5882041         326780
    Corrected Total     19    130130251


            Root MSE             571.64681    R-Square      0.9548
            Dependent Mean           16717    Adj R-Sq      0.9523
            Coeff Var            3.41951


                       Parameter Estimates

                         Parameter      Standard
    Variable      DF      Estimate        Error     t Value   Pr > |t|

    Intercept      1    -9741.92148    1362.93942       -7.15    <.0001
    radio          1     347.68885      17.83091        19.50    <.0001
```

Table 6.6 SAS Radio Regression Model: Marketing/Planning Case Study

```
                         The REG Procedure
                           Model: MODEL1
                      Dependent Variable: sales

              Number of Observations Read          20
              Number of Observations Used          20


                        Analysis of Variance

                                Sum of        Mean
     Source             DF      Squares      Square    F Value   Pr > F

     Model               1    119421443   119421443    200.73   <.0001
     Error              18     10708808      594934
     Corrected Total    19    130130251


                Root MSE           771.31951   R-Square    0.9177
                Dependent Mean        16717   Adj R-Sq    0.9131
                Coeff Var           4.61393


                        Parameter Estimates

                        Parameter     Standard
     Variable     DF     Estimate       Error    t Value   Pr > |t|

     Intercept     1       -42229    4164.12104    -10.14    <.0001
     tv            1    221.10431      15.60596     14.17    <.0001
```

Table 6.7 SAS TV Regression Model: Marketing/Planning Case Study

The computer printouts in the tables provide a variety of statistics for comparative purposes. Discussion will be limited here to just a few. The R-Square statistics are a precise proportional measure of the variation that is explained by the independent variable's behavior with the dependent variable. The closer the R-Square is to 1.00, the more of the variation is explained, and the better the predictive variable. The three variables' R-Squares are 0.0002 (POS), 0.9548 (radio), and 0.9177 (TV). Clearly, radio is the best predictor variable of the three, followed by TV and, without almost any relationship, POS. This latter result was expected based on the prior Pearson correlation. What it is suggesting is that only 0.0823 percent (1.000–0.9177) of the variation in product sales is explained by TV commercials.

From ANOVA, the F-ratio statistic is useful in actually comparing the regression model's capability to predict the dependent variable. As R-Square increases, so does the F-ratio because of the way in which they are computed and what is measured by both. The larger the F-ratio (like the R-Square statistic), the greater the statistical significance in explaining the variable's relationships. The three variables' F-ratios from the ANOVA tables are 0.00 (POS), 380.22 (radio), and 200.73 (TV). Both radio and TV are statistically significant, but POS has an insignificant relationship. To give some idea of how significant the relationships are, assuming a level of significance where $\alpha=0.01$, one would only need a cut-off value for the F-ratio of 8.10 to designate it as being significant. Not exceeding that F-ratio (as in the case of POS at 0.00) is the same as saying that the coefficient in the regression model for POS is no different from a value of zero (no contribution to Product Sales). Clearly, the independent variables radio and TV appear to have strong relationships with the dependent variable. The question is whether the two combined or even three variables might provide a more accurate forecasting model than just using the one best variable like radio.

Continuing with the step-wise multiple regression procedure, we next determine the possible combinations of variables to see if a particular combination is better than the single variable models computed previously. To measure this, we have to determine the possible combinations for the variables and compute their regression models. The combinations are (1) POS and radio; (2) POS and TV; (3) POS, radio, and TV; and (4) radio and TV.

The resulting regression model statistics are summarized and presented in Table 6.8. If one is to base the selection decision solely on the R-Square statistic, there is a tie between the POS/radio/TV and the radio/TV combination (0.979 R-Square values). If the decision is based solely on the F-ratio value from ANOVA, one would select just the radio/TV combination, which one might expect of the two most significantly correlated variables.

To aid in supporting a final decision and to ensure these analytics are the best possible estimates, we can consider an additional statistic. That tie breaker is the R-Squared (Adjusted) statistic, which is commonly used in multiple regression models.

Table 6.8 SAS Variable Combinations and Regression Model Statistics: Marketing/Planning Case Study

Variable Combination	R-Square	R-Square (Adjusted)	F-Ratio
POS/radio	0.957	0.952	188.977
POS/TV	0.920	0.911	97.662
POS/radio/TV	0.979	0.951	123.315
Radio/TV	0.979	0.953	192.555

The *R-Square Adjusted* statistic does not have the same interpretation as R-Square (a precise, proportional measure of variation in the relationship). It is instead a comparative measure of suitability of alternative independent variables. It is ideal for selection between independent variables in a multiple regression model. The R-Square adjusted seeks to take into account the phenomenon of the R-Square automatically increasing when additional independent variables are added to the model. This phenomenon is like a painter putting paint on a canvas, where more paint additively increases the value of the painting. Yet by continually adding paint, there comes a point at which some paint covers other paint, diminishing the value of the original.

Similarly, statistically adding more variables should increase the ability of the model to capture what it seeks to model. On the other hand, putting in too many variables, some of which may be poor predictors, might bring down the total predictive ability of the model. The R-Square adjusted statistic provides some information to aid in revealing this behavior.

The value of the R-Square adjusted statistic can be negative, but it will always be less than or equal to that of the R-Square in which it is related. Unlike R-Square, the R-Square adjusted increases when a new independent variable is included only if the new variable improves the R-Square more than would be expected in the absence of any independent value being added. If a set of independent variables is introduced into a regression model one at a time in forward step-wise regression using the highest correlations ordered first, the R-Square adjusted statistic will end up being equal to or less than the R-Square value of the original model. By systematic experimentation with the R-Square adjusted recomputed for each added variable or combination, the value of the R-Square adjusted will reach a maximum and then decrease. The multiple regression model with the largest R-Square adjusted statistic will be the most accurate combination of having the best fit without excessive or unnecessary independent variables. Again, just putting all the variables into a model may add unneeded variability, which can decrease its accuracy. Thinning out the variables is important.

Finally, in the step-wise multiple regression procedure, a final decision on the variables to be included in the model is needed. Basing the decision on the R-Square adjusted, the best combination is radio/TV. The SAS multiple regression model and support statistics are presented in Table 6.9.

Although there are many other additional analyses that could be performed to validate this model, we will use the SAS multiple regression model in Table 6.9 for the firm in this case study. The forecasting model can be expressed as follows:

$$Y_p = -17150 + 275.69065 X_1 + 48.34057 X_2$$

where:

Y_p = the estimated number of dollars of product sales

X_1 = the number of dollars to invest in radio commercials

X_2 = the number of dollars to invest in TV commercials

```
                           The REG Procedure
                           Model: MODEL1
                        Dependent Variable: sales

                Number of Observations Read        20
                Number of Observations Used        20

                        Analysis of Variance

                               Sum of        Mean
       Source          DF      Squares      Square    F Value   Pr > F

       Model            2    124628723    62314362     192.55   <.0001
       Error           17      5501528      323619
       Corrected Total 19    130130251

              Root MSE              568.87547   R-Square    0.9577
              Dependent Mean           16717   Adj R-Sq    0.9527
              Coeff Var              3.40294

                        Parameter Estimates

                       Parameter    Standard
       Variable    DF   Estimate      Error    t Value   Pr > |t|

       Intercept    1     -17150   6965.59100    -2.46    0.0248
       radio        1  275.69065     68.72801     4.01    0.0009
       tv           1   48.34057     44.58042     1.08    0.2934
```

Table 6.9 SAS Best Variable Combination Regression Model and Statistics: Marketing/Planning Case Study

Because all the data used in the model is expressed as dollars, the interpretation of the model is made easier than using more complex data. The interpretation of the multiple regression model suggests that for every dollar allocated to radio commercials (represented by X_1), the firm will receive $275.69 in product sales (represented by Y_p in the model). Likewise, for every dollar allocated to TV commercials (represented by X_2), the firm will receive $48.34 in product sales.

A caution should be mentioned on the results of this case study. Many factors might challenge a result, particularly those derived from using powerful and complex methodologies like multiple regression. As such, the results may not occur as estimated, because the model is not reflecting past performance. What is being suggested here is that more analysis can always be performed in questionable situations. Also, additional analysis to confirm a result should be undertaken to strengthen the trust that others must have in the results to achieve the predicted higher levels of business performance.

In summary, for this case study, the predictive analytics analysis has revealed a more detailed, quantifiable relationship between the generation of product sales and the sources of promotion that best predict sales. The best way to allocate the $350,000 budget to maximize product sales might involve placing the entire budget into radio commercials because they give the best return per dollar of budget. Unfortunately, there are constraints and limitations regarding what can be allocated to the different types of promotional methods. Optimizing the allocation of a resource and maximizing business performance necessitate the use of special business analytic methods designed to accomplish this task. This requires the additional step of prescriptive analytics analysis in the BA process, which will be presented in the last section of Chapter 7.

Summary

This chapter dealt with the predictive analytics step in the BA process. Specifically, it discussed logic-driven models based on experience and aided by methodologies like the cause-and-effect and the influence diagrams. This chapter also defined data-driven models useful in the predictive step of the BA analysis. A further discussion of data mining was presented. Data mining methodology such as neural networks, discriminant analysis, logistic regression, and hierarchical clustering was described. An illustration of K-mean clustering using SAS was presented. Finally, this chapter discussed the second installment of a case study illustrating the predictive analytics step of the BA process. The remaining installment of the case study will be presented in Chapter 7.

Once again, several of this book's appendixes are designed to augment the chapter material by including technical, mathematical, and statistical tools. For both a greater understanding of the methodologies discussed in this chapter and a basic review of statistical and other quantitative methods, a review of the appendixes is recommended.

As previously stated, the goal of using predictive analytics is to generate a forecast or path for future improved business performance. Given this predicted path, the question now is how to exploit it as fully as possible. The purpose of the prescriptive analytics step in the BA process is to serve as a guide to fully maximize the outcome in using the information provided by the predictive analytics step. The subject of Chapter 7 is the prescriptive analytics step in the BA process.

Discussion Questions

1. Why is predictive analytics analysis the next logical step in any business analytics (BA) process?

2. Why would one use logic-driven models to aid in developing data-driven models?

3. How are neural networks helpful in determining both associations and classification tasks required in some BA analyses?

4. Why is establishing clusters important in BA?

5. Why is establishing associations important in BA?

6. How can F-tests from the ANOVA be useful in BA?

Problems

1. Using a similar equation to the one developed in this chapter for predicting dollar product sales (note below), what is the forecast for dollar product sales if the firm could invest $70,000 in radio commercials and $250,000 in TV commercials?

$$Y_p = -17150.455 + 275.691 X_1 + 48.341 X_2$$

where:

Y_p = the estimated number of dollars of product sales

X_1 = the number of dollars to invest in radio commercials

X_2 = the number of dollars to invest in TV commercials

2. Using the same formula as in Question 1, but now using an investment of $100,000 in radio commercials and $300,000 in TV commercials, what is the prediction on dollar product sales?

3. Assume for this problem the following table would have held true for the resulting marketing/planning case study problem. Which combination of variables is estimated here to be the best predictor set? Explain why.

Variable Combination	R-Square	R-Square (Adjusted)	F-Ratio
POS/radio	0.057	0.009	2.977
POS/TV	0.120	0.100	3.662
POS/radio/TV	0.179	0.101	4.315
Radio/TV	0.879	0.853	122.555

4. Assume for this problem that the following table would have held true for the resulting marketing/planning case study problem. Which of the variables is estimated here to be the best predictor? Explain why.

Statistic	Radio	Paper	TV	POS
Pearson Correlation r with product sales	.127	.083	.208	.013
Significance Test (1-Tailed)	.212	.313	.192	.479

References

Abrahams, A. S., Jiao, J., Fan, W., Wang, G., Zhang, Z. (2013). "What's Buzzing in the Blizzard of Buzz? Automotive Component Isolation in Social Media Postings." *Decision Support Systems*. Vol. 55, No. 4, pp. 871–882.

Evans, J. R. (2013). *Business Analytics*. Pearson Education, Upper Saddle River, NJ.

Nisbet, R., Elder, J., Miner, G. (2009). *Handbook of Statistical Analysis & Data Mining Applications*. Academic Press, Burlington, MA.

Schniederjans, M. J., Cao, Q., Triche, J. H. (2014). *E-Commerce Operations Management*, 2nd ed. World Scientific, Singapore.

7

What Is Prescriptive Analytics?

Chapter objectives:

- List and describe the commonly used prescriptive analytics in the business analytics (BA) process.
- Explain the role of case studies in prescriptive analytics.
- Explain how fitting a curve can be used in prescriptive analytics.
- Explain how to formulate a linear programming model.
- Explain the value of linear programming in the prescriptive analytics step of BA.

7.1 Introduction

After undertaking the descriptive and predictive analytics steps in the BA process, one should be positioned to undertake the final step: prescriptive analytics analysis. The prior analysis should provide a forecast or prediction of what future trends in the business may hold. For example, there may be significant statistical measures of increased (or decreased) sales, profitability trends accurately measured in dollars for new market opportunities, or measured cost savings from a future joint venture.

If a firm knows where the future lies by forecasting trends, it can best plan to take advantage of possible opportunities that the trends may offer. Step 3 of the BA process, prescriptive analytics, involves the application of decision science, management science, or operations research methodologies to make best use of allocable resources. These are mathematically based methodologies and algorithms designed to take variables and other parameters into a quantitative framework and generate an optimal or near-optimal solution to complex problems. These methodologies can be used to optimally allocate a firm's limited resources to take best advantage of the opportunities it has found in the predicted future trends. Limits on human, technology, and financial resources prevent any firm from going after all the opportunities. Using prescriptive analytics allows the firm to allocate limited resources to optimally or near-optimally achieve the objectives as fully as possible.

In Chapter 3, "What Resource Considerations Are Important to Support Business Analytics?" the relationships of methodologies to the BA process were expressed as a function of certification exam content. The listing of the prescriptive analytic methodologies as they are in some cases utilized in the BA process is again presented in Figure 7.1 to form the basis of this chapter's content.

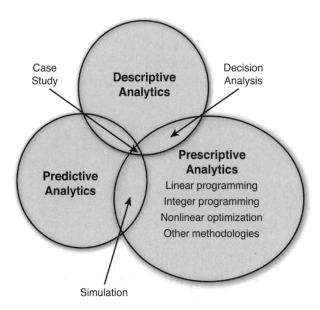

Figure 7.1 Prescriptive analytic methodologies

7.2 Prescriptive Modeling

The listing of prescriptive analytic methods and models in Figure 7.1 is but a small grouping of many operations research, decision science, and management science methodologies that are applied in this step of the BA process. Most of the methodologies in Table 7.1 are explained throughout this book. (See the Additional Information column in Table 7.1.)

Table 7.1 Select Prescriptive Analytic Models

Data-Driven Models	Possible Applications	Additional Information
Linear Programming (LP)	A general-purpose modeling methodology is applied to multiconstrained, multivariable problems when an optimal solution is sought. It is ideal for complex and large-scale problems when limited resources are being allocated to multiple uses. Examples include allocating advertising budgets to differing media, allocating human and technology resources to product production, and optimizing blends of mixing ingredients to minimize costs of food products.	Chapters 7 and 8, "A Final Business Analytics Case Problem" Appendix B, "Linear Programming" Appendix C, "Duality and Sensitivity Analysis in Linear Programming"
Integer Programming	This is the same as LP, but it permits decision variables to be integer values. Examples include allocating stocks to portfolios, allocating personnel to jobs, and allocating types of crops to farm lands.	Appendix D, "Integer Programming"
Nonlinear Optimization	A large class of methodologies and algorithms is used to analyze and solve for optimal or near-optimal solutions when the behavior of the data is nonlinear. Examples include solving for optimized allocations of human, technology, and systems whose data appears to form a cost or profit function that is quadratic, cubic, or nonlinear in some way.	Chapters 7 and 8 Appendix E, "Forecasting"
Decision Analysis	A set of methodologies, models, or principles is used to analyze and guide decision-making when multiple choices face the decision maker in differing decision environments (for example, certainty, risk, and uncertainty). Examples include selecting one from a set of computer systems, trucks, or site locations for a service facility.	Appendix G, "Decision Theory"
Case Studies	A learning aid provides practical experience by offering real or hypothetical case studies of real-world applications of BA. For example, case studies can simulate the issues and challenges in an actual problem setting. This kind of simulation can prep decision makers to anticipate and prepare for what has been predicted to occur by the predicted analytics step in the BA process. For example, a case study discussion on how to cope with organization growth might provide a useful decision-making environment for a firm whose analytics have predicted growth in the near future.	This is beyond the scope of this book. See Sekaran and Bougie (2013); Adkins (2006).

Data-Driven Models	Possible Applications	Additional Information
Simulation	This methodology can be used in prescriptive analysis in situations where parameters are probabilistic, nonlinear, or just too complex to use with other optimization models that require deterministic or linear behavior. For example, a bank might want to simulate the transactions it currently uses to process a loan application to determine if changes in the process might reduce time and improve performance. The simulation model might be used to test alternative process scenarios.	Appendix F, "Simulation"
Other Methodologies	The areas of operations research, decision sciences, and management science combine the application of mathematics, engineering, and computer science to offer a broad listing of prescriptive methodologies. These other methodologies include network modeling, project scheduling, dynamic programming, queuing models, decision support systems, heuristics, artificial intelligence, expert systems, Markov processes, decision tree analysis, game theory, goal programming, nonlinear programming, reliability analysis, genetic programming, and data envelopment analysis, just to name a few. There are virtually no application limitations on the collection of these methodologies.	These are outside the scope of this book. See Hillier (2014); Cooper et al. (2013); Rothlauf (2013); Liebowitz (2014); Albright and Winston (2014).

7.3 Nonlinear Optimization

The prescriptive methodologies in Table 7.1 are explained in detail in the referenced chapters and appendixes, but nonlinear optimization will be discussed here. When business performance cost or profit functions become too complex for simple linear models to be useful, exploration of nonlinear functions is a standard practice in BA. Although the predictive nature of exploring for a mathematical expression to denote a trend or establish a forecast falls mainly in the predictive analytics step of BA, the use of the nonlinear function to optimize a decision can fall in the prescriptive analytics step.

As mentioned previously, there are many mathematical programing nonlinear methodologies and solution procedures designed to generate optimal business performance solutions. Most of them require careful estimation of parameters that may or may not be accurate, particularly given the precision required of a solution that can be so precariously dependent upon parameter accuracy. This precision is further complicated in BA by the large data files that should be factored into the model-building effort.

To overcome these limitations and be more inclusive in the use of large data, we can apply regression software. As illustrated in Appendix E, curve-fitting software can be used to generate predictive analytic models that can also be utilized to aid in making prescriptive analytic decisions.

For purposes of illustration, SAS's software will be used to fit data to curves in this chapter. Suppose that a resource allocation decision is being faced whereby one must decide how many computer servers a service facility should purchase to optimize the firm's costs of running the facility. The firm's predictive analytics effort has shown a growth trend. A new facility is called for if costs can be minimized. The firm has a history of setting up large and small service facilities and has collected the 20 data points in Figure 7.2. Whether there are 20 or 20,000 items in the data file, SAS can be used to fit data based on regression mathematics to a nonlinear line that best minimizes the distance from the data items to the line. The software then converts the line into a mathematical expression useful for forecasting.

	server	cost
1	1	27654
2	2	24789
3	3	21890
4	4	21633
5	5	15843
6	6	12567
7	7	8943
8	8	6789
9	9	4533
10	10	4678
11	11	5321
12	12	5765
13	13	5432
14	14	9995
15	15	13522
16	16	17563
17	17	22732
18	18	22643
19	19	24621
20	20	28111

Figure 7.2 Data for SAS curve fitting

In this server problem, the basic data has a u-shaped function, as presented in Figure 7.3. This is a classic shape for most cost functions in business. In this problem, it represents the balancing of having too few servers (resulting in a costly loss of customer business through dissatisfaction and complaints with the service) or too many servers (excessive waste in investment costs because of underutilized servers). Although this is an overly simplified example with little and nicely ordered data for clarity purposes, in big data situations, cost functions are considerably less obvious.

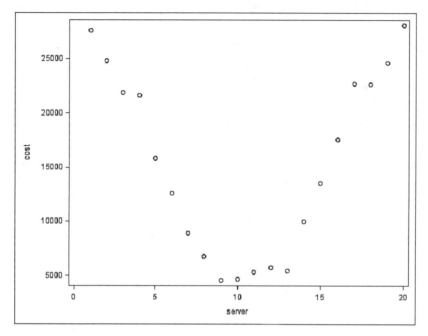

Figure 7.3 Server problem basic data cost function

The first step in curve fitting is to generate the best-fitting curve to the data. Using SAS and the data in Figure 7.2, the regression process seeks to minimize the distance by creating a line in one of the eight regression models in Figure 7.3. Doing this in SAS requires the selection of a set of functions that the analyst might believe is a good fit. The number of regression functions selected can be flexible. SAS offers a wide number of possible regression models to choose from. The result is a series of regression models and statistics, including ANOVA and other testing statistics. It is known from the previous illustration of regression that the adjusted R-Square statistic can reveal the best estimated relationship between the independent (number of servers) and dependent (total cost) variables. These statistics are presented in Table 7.2. The best adjusted R-Square value (the largest) occurs with the quadratic model, followed by the cubic model. The more detailed supporting statistics for both of these models are presented in Table 7.3. The graph for all the SPSS curve-fitting models appears in Figure 7.4.

Linear

Root MSE	8687.28965	R-Square	0.0011
Dependent Mean	15251	Adj R-Sq	-0.0544
Coeff Var	56.96135		

Logarithmic

Root MSE	8376.01994	R-Square	0.0714
Dependent Mean	15251	Adj R-Sq	0.0198
Coeff Var	54.92040		

Inverse

Root MSE	7825.69613	R-Square	0.1894
Dependent Mean	15251	Adj R-Sq	0.1444
Coeff Var	51.31200		

Quadratic

Root MSE	2342.31463	R-Square	0.9314
Dependent Mean	15251	Adj R-Sq	0.9233
Coeff Var	15.35823		

Cubic

Root MSE	2404.00949	R-Square	0.9320
Dependent Mean	15251	Adj R-Sq	0.9193
Coeff Var	15.76276		

S-Curve

Root MSE	0.62652	R-Square	0.1454
Dependent Mean	9.44856	Adj R-Sq	0.0979
Coeff Var	6.63085		

Logistic

R-Square	0.0047	Max-rescaled R-Square	0.0047

Growth

Root MSE	0.67750	R-Square	0.0006
Dependent Mean	9.44856	Adj R-Sq	-0.0549
Coeff Var	7.17041		

Table 7.2 Adjusted R-Square Values of All SAS Models

Quadratic-Full

```
                        The REG Procedure
                          Model: MODEL1
                      Dependent Variable: cost

         Number of Observations Read        20
         Number of Observations Used        20
```

Analysis of Variance

Source	DF	Sum of Squares	Mean Square	F Value	Pr > F
Model	2	1266704838	633352419	115.44	<.0001
Error	17	93269443	5486438		
Corrected Total	19	1359974281			

Root MSE	2342.31463	R-Square	0.9314	
Dependent Mean	15251	Adj R-Sq	0.9233	
Coeff Var	15.35823			

Parameter Estimates

Variable	DF	Parameter Estimate	Standard Error	t Value	Pr > |t|
Intercept	1	35418	1742.63922	20.32	<.0001
server	1	-5589.43151	382.18778	-14.62	<.0001
quadratic_server	1	268.44919	17.67797	15.19	<.0001

Cubic-Full

```
                        The REG Procedure
                          Model: MODEL1
                      Dependent Variable: cost

         Number of Observations Read        20
         Number of Observations Used        20
```

Analysis of Variance

Source	DF	Sum of Squares	Mean Square	F Value	Pr > F
Model	3	1267506095	422502032	73.11	<.0001
Error	16	92468186	5779262		
Corrected Total	19	1359974281			

Root MSE	2404.00949	R-Square	0.9320	
Dependent Mean	15251	Adj R-Sq	0.9193	
Coeff Var	15.76276			

Parameter Estimates

Variable	DF	Parameter Estimate	Standard Error	t Value	Pr > |t|
Intercept	1	36134	2625.97612	13.76	<.0001
server	1	-5954.73759	1056.59555	-5.64	<.0001
quadratic_server	1	310.89531	115.43050	2.69	0.0160
cubic_server	1	-1.34750	3.61891	-0.37	0.7145

Table 7.3 Quadratic and Cubic Model SAS Statistics

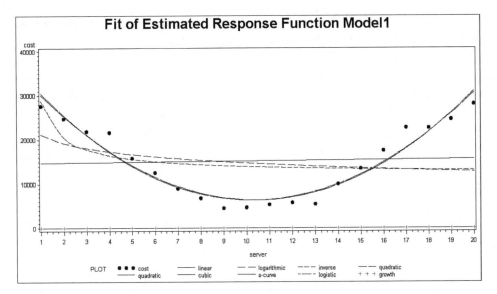

Figure 7.4 Graph of all SAS curve-fitting models

From Table 7.3, the resulting two statistically significant curve-fitted models follow:

$$Y_p = 35418 - 5589.432\ X + 268.445\ X^2\ \text{[Quadratic model]}$$
$$Y_p = 36134 - 5954.738\ X + 310.895\ X^2 - 1.347\ X^3\ \text{[Cubic model]}$$

where:

Y_p = the forecasted or predicted total cost
X = the number of computer servers

For purposes of illustration, we will use the quadratic model. In the next step of using the curve-fitted models, one can either use calculus to derive the cost minimizing value for X (number of servers) or perform a deterministic simulation where values of X are substituted into the model to compute and predict the total cost (Y_p). The calculus-based approach is presented in the "Addendum" section of this chapter.

As a simpler solution method to finding the optimal number of servers, simulation can be used. Representing a deterministic simulation (see Appendix F, Section F.2.1), the resulting costs of servers can be computed using the quadratic model, as presented in Figure 7.5. These values were computed by plugging the number of server values (1 to 20) into the Y_p quadratic function one at a time to generate the predicted values for each of the server possibilities. Note that the lowest value in these predicted values occurs with the acquisition of 10 servers at $6367.952, and the next lowest is at 11 servers at $6415.865. In the actual data in Figure 7.2, the minimum total cost point occurs at 9 servers at $4533, whereas the next lowest total cost is $4678 occurring at 10 servers. The differences are due to the estimation process of curve fitting. Note in Figure 7.3 that the curve that is fitted does not touch the lowest 5 cost values. Like regression in general, it is an estimation process, and although the ANOVA statistics in the quadratic model demonstrate a strong relationship with the actual values, there is some error. This process provides a near-optimal solution but does not guarantee one.

	A	G
1	1	30096.79
2	2	25312.69
3	3	21065.48
4	4	17355.16
5	5	14181.74
6	6	11545.2
7	7	9445.553
8	8	7882.796
9	9	6856.929
10	10	6367.952
11	11	6415.865
12	12	7000.668
13	13	8122.361
14	14	9780.944
15	15	11976.42
16	16	14708.78
17	17	17978.03
18	18	21784.18
19	19	26127.21
20	20	31007.13

Figure 7.5 Predicted total cost in server problem for each server alternative

Like all regression models, curve fitting is an estimation process with risks, but the supporting statistics, like ANOVA, provide some degree of confidence in the resulting solution.

Finally, it must be mentioned that many other nonlinear optimization methodologies exist. Some, like quadratic programming, are considered constrained optimization models (like LP). These topics are beyond the scope of this book. For additional information on nonlinear programming, see King and Wallace (2013), Betts (2009), and Williams (2013). Other methodologies, like the use of calculus in this chapter, are useful in solving for optimal solutions in unconstrained problem settings. For additional information on calculus methods, see Spillers and MacBain (2009), Luptacik (2010), and Kwak and Schniederjans (1987).

7.4 Continuation of Marketing/Planning Case Study Example: Prescriptive Step in the BA Analysis

In Chapter 5, "What Is Descriptive Analytics?" and Chapter 6, "What Is Predictive Analytics?" an ongoing marketing/planning case study was presented to illustrate some of the tools and strategies used in a BA problem analysis. This is the third and final installment of the case study dealing with the prescriptive analytics step in BA.

7.4.1 Case Background Review

The predictive analytics analysis in Chapter 6 revealed a statistically strong relationship between radio and TV commercials that might be useful in predicting future product sales. The ramifications of these results suggest a better allocation of funds away from paper and POS ads to radio and TV commercials. Determining how much of the $350,000 budget should be allocated between the two types of commercials requires the application of an optimization decision-making methodology.

7.4.2 Prescriptive Analysis

The allocation problem of the budget to purchase radio and TV commercials is a multivariable (there are two media to consider), constrained (there are some limitations on how one can allocate the budget funds), optimization problem (BA always seeks to optimize business performance). Many optimization methods could be employed to determine a solution to this problem. Considering the singular objective of maximizing estimated product sales, linear programming (LP) is an ideal methodology to apply in this situation. To employ LP to model this problem, use the six-step LP formulation procedure explained in Appendix B.

7.4.2.1 Formulation of LP Marketing/Planning Model

In the process of exploring the allocation options, a number of limitations or constraints on placing radio and TV commercials were observed. The total budget for all the commercials was set at a maximum of $350,000 for the next monthly campaign. To receive the radio commercial price discount requires a minimum budget investment of $15,000. To receive the TV commercials price discount requires a minimum budget investment of $75,000. Because the radio and TV stations are owned by the same corporation, there is an agreement that for every dollar of radio commercials required, the client firm must purchase $2 in TV commercials. Given these limitations and the modeled relationship found in the previous predictive analysis, one can for-mulate the budget allocation decision as an LP model using a five-step LP formulation procedure (see Appendix B, Section B.4.1):

1. **Determine the type of problem**—This problem seeks to maximize dollar product sales by determining how to allocate budget dollars over radio and TV commercials. For each dollar of radio commercials estimated with the regres-sion model, \$275.691 will be received, and for each dollar of TV commercials, \$48.341 will be received. Those two parameters are the product sales values to maximize. Therefore, it will be a maximization model.

2. **Define the decision variables**—The decision variables for the LP model are derived from the multiple regression model's independent variables. The only adjustment is the monthly timeliness of the allocation of the budget:

   X_1 = the number of dollars to invest in radio commercials for the next monthly campaign

   X_2 = the number of dollars to invest in TV commercials for the next monthly campaign

3. **Formulate the objective function**—Because the multiple regression model defines the dollar sales as a linear function with the two independent variables, the same dollar coefficients from the regression model can be used as the con-tribution coefficients in the objective function. This results in the following LP model objective function:

   Maximize: $Z = 275.691 \, X_1 + 48.341 \, X_2$

4. **Formulate the constraints**—Given the information on the limitations in this problem, there are four constraints:

   Constraint 1—No more than \$350,000 is allowed for the total budget to allo-cate to both radio (X_1) and TV (X_2) commercials. So add $X_1 + X_2$ and set it less than or equal to 350,000 to formulate the first constraint as follows:

   $X_1 + X_2 \leq 350000$

   Constraint 2—To get a discount on radio (X_1) commercials, the firm must allo-cate a minimum of \$15,000 to radio. The constraint for this limitation follows:

   $X_1 \geq 15000$

   Constraint 3—Similar to Constraint 2, to get a discount on TV (X_2) commer-cials, the firm must allocate a minimum of \$75,000 to TV. The constraint for this limitation follows:

   $X_2 \geq 75000$

Constraint 4—This is a blending problem constraint (see Appendix B, Section B.6.3). What is needed is to express the relationship as follows:

$$\frac{X_1}{1} = \frac{X_2}{2}$$

which is to say, for each one unit of X_1, one must acquire two units of X_2. Said differently, the ratio of one unit of X_1 is equal to two units of X_2. Given the expression, use algebra to cross-multiply such that

$$2\,X_1 = X_2$$

Convert it into an acceptable constraint with a constant on the right side and the variables on the left side as follows:

$$2\,X_1 - X_2 = 0$$

5. **State the nonnegativity and given requirements**—With only two variables, this formal requirement in the formulation of an LP model is expressed as follows:

$$X_1, X_2 \geq 0$$

Because these variables are in dollars, they do not have to be integer values. (They can be any real or cardinal number.) The complete LP model formulation is given here:

Maximize: $Z = 275.691\,X_1 + 48.341\,X_2$
Subject to: $X_1 + X_2 \leq 350000$
$\qquad\qquad X_1 \qquad \geq 15000$
$\qquad\qquad\qquad X_2 \geq 75000$
$\qquad 2\,X_1 - X_2 = 0$
and $\qquad X_1, X_2 \geq 0$

7.4.2.2 Solution for the LP Marketing/Planning Model

Appendix B explains that both Excel and LINGO software can be used to run the LP model and solve the budget allocation in this marketing/planning case study problem. For purposes of brevity, discussion will be limited to just LINGO. As will be presented in Appendix B, *LINGO* is a mathematical programming language and software system. It allows the fairly simple statement of the LP model to be entered into a single window and run to generate LP solutions.

LINGO opens with a blank window for entering whatever type of model is desired. After the LP model formulation is entered into the LINGO software, the resulting data entry information is presented in Figure 7.6.

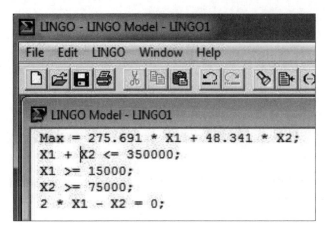

Figure 7.6 LINGO LP model entry requirements: marketing/planning case study

There are several minor differences in the model entry requirements over the usual LP model formulation. These differences are required to run a model in LINGO. These include (1) using the term "Max" instead of "Maximize," (2) dropping off "Subject to" and "and" in the model formulation, (3) placing an asterisk and a space between unknowns and constant values in the objective and constraint functions where multiplication is required, (4) ending each expression with a semicolon, and (5) omitting the nonnegativity requirements, which aren't necessary.

Now that the model is entered into LINGO, a single click on the SOLVE option in the bar at the top of the window generates a solution. The marketing budget allocation LP model solution is found in Figure 7.7.

As it turns out, the optimal distribution of the $350,000 promotion budget is to allocate $116,666.70 to radio commercials and $233,333.30 to TV commercials. The resulting Z value, which in this model is the total predicted product sales in dollars, is 0.4344352E+08, or $43,443,524. When we compare that future estimated month's product sales with the average current monthly product sales of $16,717,200 presented in Figure 7.7, it does appear that the firm in this case study will optimally maximize future estimated monthly product sales if it allocates the budget accordingly (that is, if the multiple regression model estimates and the other parameters in the LP model hold accurate and true).

```
 Solution Report - LINGO1
  Global optimal solution found.
  Objective value:                              0.4344352E+08
  Total solver iterations:                          0


              Variable           Value        Reduced Cost
                    X1        116666.7            0.000000
                    X2        233333.3            0.000000

                   Row  Slack or Surplus          Dual Price
                     1     0.4344352E+08            1.000000
                     2         0.000000          124.1243
                     3        101666.7            0.000000
                     4        158333.3            0.000000
                     5         0.000000           75.78333
```

Figure 7.7 LINGO LP model solution: marketing/planning case study

In summary, the prescriptive analytics analysis step brings the prior statistical analytic steps into an applied decision-making process where a potential business performance improvement is shown to better this organization's ability to use its resources more effectively. The management job of monitoring performance and checking to see that business performance is in fact improved is a needed final step in the BA analysis. Without proof that business performance is improved, it's unlikely that BA would continue to be used.

7.4.2.3 Final Comment on the Marketing/Planning Model

Although the LP solution methodology used to generate an allocation solution guarantees an optimal LP solution, it does not guarantee that the firm using this model's solution will achieve the results suggested in the analysis. Like any forecasting estimation process, the numbers are only predictions, not assurances of outcomes. The high levels of significance in the statistical analysis and the added use of other conformational statistics (R-Square, adjusted R-Square, ANOVA, and so on) in the model development provide some assurance of predictive validity. There are many other methods and approaches that could have been used in this case study. Learning how to use more statistical and decision science tools helps ensure a better solution in the final analysis.

Summary

This chapter discussed the prescriptive analytics step in the BA process. Specifically, this chapter revisited and briefly discussed methodologies suggested in BA certification exams. An illustration of nonlinear optimization was presented to demonstrate how the combination of software and mathematics can generate useful decision-making information. Finally, this chapter presented the third installment of a marketing/planning case study illustrating how prescriptive analytics can benefit the BA process.

We end this book with a final application of the BA process. Once again, several of the appendixes are designed to augment this chapter's content by including technical, mathematical, and statistical tools. For both a greater understanding of the methodologies discussed in this chapter and a basic review of statistical and other quantitative methods, a review of the appendixes and chapters is recommended.

Addendum

The *differential calculus* method for finding the minimum cost point on the quadratic function that follows involves a couple of steps. It finds the zero slope point on the cost function (the point at the bottom of the u-shaped curve where a line could be drawn that would have a zero slope). There are limitations to its use, and qualifying conditions are required to prove minimum or maximum positions on a curve. The quadratic model in the server problem follows:

$Y_p = 35418 - 5589.432\,X + 268.445\,X^2$ [Quadratic model]

Step 1. Given the quadratic function above, take its first derivative.

$d(Y_p) = -5589.432 + 536.89\,X$

Step 2. Set the derivative function equal to zero and solve for X.

$0 = -5589.432 + 536.89\,X$

$X = 10.410758$

Slightly more than ten servers should be purchased at the resulting optimally minimized cost value. This approach provides a near-optimal solution but does not guarantee one. For additional information on the application of calculus, see Field (2012) and Dobrushkin (2014).

Discussion Questions

1. How are prescriptive and descriptive analytics related?
2. How can we use simulation in both predictive and prescriptive analytics?
3. Why in the server problem were there so few statistically significant models?
4. Does it make sense that the resulting quadratic model in Figure 7.4 did not touch the lowest cost data points in the data file? Explain.
5. What conditions allowed the application of LP?

Problems

1. A computer services company sells computer services to industrial users. The company's analytics officer has predicted the need for growth to meet competitive pressures. To implement this strategy, upper management has determined that the company would tactically expand its sales and service organization. In this expansion, new districts would be defined and newly hired or appointed managers would be placed in charge to establish and run the new districts. The first job of the new district managers would be to select the salespeople and staff support employees for their districts. To aid the new district managers in deciding on the number of salespeople and staffers to hire, the company researched existing office operations and made a number of analytic-based observations, which it passed on to the new district managers. A new manager's district should, at the very least, have 14 salespeople and 4 staffers to achieve adequate customer service. Research has indicated that a district manager could adequately manage the equivalent of no more than 32 employees. Salespeople are twice as time consuming to manage as staffers. The district manager was assigned part of the floor in an office building for operations. This space could house no more than 20 salespeople and staffers. The district manager had some discretion regarding budgetary limitations. A total payroll budget for salespeople and staffers was set at $600,000. The manufacturing company's policy in developing a new territory would be to pay salespeople a fixed salary instead of commissions and salary. The yearly salary of a beginning salesperson would be $36,000, whereas a staffer would receive $18,000. All the salespeople and staffers being hired for this district would be new with the company, and as such, would start with the basic salaries mentioned. Finally, the source of prospective salespeople and staffers would be virtually unlimited in the district and pose no constraint on the problem situation. What is the LP formulation of this model?

2. (This problem requires computer support.) What is the optimal answer to the problem formulated in Problem 1?

3. A trucking firm must transport exactly 900, 800, 700, and 1,000 units of a product to four cities: A, B, C, and D. The product is manufactured and supplied in two other cities, X and Y, in the exact amounts to match the total demand. The production of units from the two cities is 1,900 and 1,500 units, respectively, to X and Y. The cost per unit to transport the product between the manufacturing plants in cities X and Y and the demand market cities A, B, C, and D is given here:

Supply Plant	Demand Market			
	A	B	C	D
X	.65	.70	.80	.90
Y	.60	.60	.80	.70

For example, in the table, $0.65 is the cost to ship one unit from Supply Plant X to Demand Market A. The trucking firm needs to know how many units should be shipped from each supply city to each demand city in such a way that it minimizes total costs. Hint: This is a multidimensional decision variable problem (see Section B.6.4 in Appendix B). What is the LP model formulation for this problem?

4. (This problem requires computer support.) What is the optimal answer to the problem formulated in Problem 3?

References

Adkins, T. C. (2006). *Case Studies in Performance Management: A Guide from the Experts*. Wiley, New York, NY.

Albright, S. C., Winston, W. L. (2014). *Business Analytics: Data Analysis & Decision Making*. Cengage Learning, Stamford, CT.

Betts, J. T. (2009). *Practical Methods for Optimal Control and Estimation Using Nonlinear Programming*, 2nd ed., Society for Industrial & Applied Mathematics, London.

Cooper, W. W., Seiford, L. M., Zhu, J. (2013). *Handbook on Data Envelopment Analysis*. Springer, New York, NY.

Dobrushkin, V. A. (2014). *Applied Differential Equations: An Introduction*. Chapman and Hall/CRC, New York, NY.

Field, M. J. (2012). *Differential Calculus and Its Applications*. Dover Publishing, Mineola, NY.

Hillier, F. S. (2014). *Introduction to Operations Research*, 10th ed., McGraw-Hill Higher Education, Boston, MA.

King, A. J., Wallace, S. W. (2013). *Modeling with Stochastic Programming*, Springer, New York, NY.

Kwak, N. K, Schniederjans, M. J. (1987). *Introduction to Mathematical Programming*. Kreiger Publishing, Malabar, FL.

Liebowitz, J. (2014). *Business Analytics: An Introduction*. Auerbach Publications, New York, NY.

Luptacik, M. (2010). *Mathematical Optimization and Economic Analysis*. Springer, New York, NY.

Rothlauf, F. (2013). *Design and Modern Heuristics: Principles and Application*. Springer, New York, NY.

Sekaran, U., Bougie, R. (2013). *Research Methods for Business: A Skill-Building Approach*. Wiley, New York, NY.

Spillers, W. R., MacBain, K. M. (2009). *Structural Optimization*. Springer, New York, NY.

Williams, H. P. (2013). *Modeling Building in Mathematical Programming*. Wiley, New York, NY.

8

A Final Business Analytics Case Problem

Chapter objectives:

- Provide a capstone business analytics (BA) overview within a case study problem.
- Show the step-wise connections of the descriptive, predictive, and prescriptive steps in the BA process.

8.1 Introduction

In Parts I, "What Is Business Analytics?" and II, "Why Is Business Analytics Important?" (Chapters 1 through 3), this book explained what BA is about and why it is important to business organization decision-making. In Part III, "How Can Business Analytics Be Applied?" (Chapters 4 through 7), we explained and illustrated how BA can be applied using a variety of concepts and methodologies. Completing Part III, we seek in this chapter a closing illustration of how the BA process can be applied by presenting a final case study. This case study is meant as a capstone learning experience on the business analytics process discussed throughout the book. Several of the concepts and methodologies presented in prior chapters and the appendixes will once again be applied here.

As will be seen in this case study, unique metrics and measures are sometimes needed in a BA setting to effect a solution to a problem or answer a question. Therefore, the methodologies and approach used in this chapter should be viewed as just one approach in obtaining the desired information.

Undertaking the analytic steps in the BA process (see Chapter 1, "What Is Business Analytics?") requires a beginning effort that preempts data collection efforts. This prerequisite to BA is to understand the business systems that are part of the problem. When BA effort has been outsourced (see Chapter 4, "How Do We Align Resources to Support Business Analytics within an Organization?") or when it is completely performed in-house by a BA team (Chapter 3, "What Resource Considerations Are Important to Support Business Analytics?"), experienced managers must be brought into the process to provide the necessary systems behavior and general knowledge of operations needed to eventually model and explain how the business operates. In this case study, it is assumed that the staff or information is available. Based on this information, a BA project can be undertaken.

8.2 Case Study: Problem Background and Data

A Midwest U.S. commercial manufacturing firm is facing a supply chain problem. The manufacturer produces and sells a single product, a general-purpose small motor as a component part to different customers who incorporate the motor into their various finished products. The manufacturer has a supply chain network that connects production centers located in St. Louis, Missouri, and Dallas, Texas, with six warehouse facilities that serve commercial customers located in Kansas City, Missouri; Chicago, Illinois; Houston, Texas; Oklahoma City, Oklahoma; Omaha, Nebraska; and Little Rock, Arkansas.

Part of the supply chain problem is the need to keep the cost of shipping motors to the customers as low as possible. The manufacturer adopted a lean management philosophy that seeks to match what it produces with what is demanded at each warehouse. The problem with implementing this philosophy is complicated by the inability to forecast the customer demand month to month. If the forecast of customer demand is too low and not enough inventory is available (an underage of inventory), the manufacturer has to rush order motors that end up being costly to the manufacturer. If the forecast is too high and the manufacturer produces and ships unwanted inventory (an overage of inventory), the warehouse incurs wasteful storage costs. The management of the manufacturing firm has decided that an analytics-based procedure needs to be developed to improve overall business performance. Analysts could use this procedure each month to develop an optimal supply chain schedule of shipments from the two supply centers to the six warehouse demand destinations that would minimize costs. A key part of this procedure would be to include a means to accurately forecast customer demand and an optimization process for shipping products from the manufacturing centers to the warehouse demand destinations.

The manufacturing firm created a small BA team to develop the procedure (see Chapter 4, Section 4.1.1). The BA team consists of a BA analyst (who would be responsible for using the procedure and heading the BA team), the supply chain general manager, the shipping manager (responsible for drafting the shipping schedule), and a warehouse manager (whose job it is to develop monthly forecasts).

8.3 Descriptive Analytics Analysis

Determining a procedure by which analyst teams can determine optimal shipments between supply sources and demand destinations requires differing types of data. There is supply, demand, and cost data required to plan shipments. The total manufactured supply of motors produced at the St. Louis and Dallas plants is determined once the forecast demand is established. The BA team established that there is ample capacity between both plants to satisfy the forecasted customer demand at the six warehouse demand destinations.

The BA team determined that the cost data for shipping a motor from the production centers to the customers depends largely on distance between the cities, where the items are trucked directly by the manufacturer to the warehouses. The cost data per motor shipped to a customer is given in Table 8.1. For example, it costs the manufacturer $4 per motor to ship from St. Louis to Kansas City. These cost values are routinely computed by the manufacturer's cost accounting department and are assumed by the BA team to be accurate.

Table 8.1 Estimated Shipping Costs Per Motor

Supply Center	Kansas City, Missouri	Chicago, Illinois	Houston, Texas	Oklahoma City, Oklahoma	Omaha, Nebraska	Little Rock, Arkansas
St. Louis, Missouri	$4	$6	$9	$8	$5	$6
Dallas, Texas	$5	$8	$2	$5	$8	$5

The present system of forecasting customer demand usually results in costly overages and underages shipped to the warehouses. In the past, the manufacturer would take a three-value smoothing average to estimate the monthly demand. (See Section E.6.1 in Appendix E, "Forecasting.") This evolved by taking the last three months of actual customer motor demand and averaging them to produce a forecast for the next month. The process was repeated each month for each of the six warehouses. Not making products available when customers demanded them caused lost sales, so the manufacturer would rush and ship products to customers at a loss. On the other hand, producing too much inventory meant needless production, inventory, and shipping costs.

To deal with the variability in customer demand forecasting, the manufacturer would need to develop models for each warehouse's customer demand. The customer demand data on which to build the models was collected from prior monthly demand in motors. To determine which data to include in a final sample and which to exclude, the company adopted a few simple rules to eliminate potentially useless and out-of-date data. Going back more than 27 months invited cyclical variations caused by changes in the economy that were no longer present, so that data was removed. Unfortunately, some of the data files were incomplete and required cleansing (see Chapter 4). The resulting time series data collected on warehouse customer monthly demand files is presented in Table 8.2. It was decided that the most recent three months (bold months of 25, 26, and 27) would not be included in the model development but instead would be used for validation purposes to confirm the forecasting accuracy of the resulting models. This is similar to what was referred to as a training data set and a validation data set (see Section 6.3.1 in Chapter 6, "What Is Predictive Analytics?").

Table 8.2 Actual Monthly Customer Demand in Motors

Month	Kansas City	Chicago	Houston	Oklahoma City	Omaha	Little Rock
1	3120	2130	3945	14020	5045	4610
2	3090	2290	4000	13890	5030	4630
3	3140	2405	4105	13785	5075	4650
4	3010	2580	4300	13575	5015	4680
5	2900	2635	4255	13345	5015	4700
6	2990	2690	4420	12990	5020	4750
7	3000	2740	4540	12340	5025	4800
8	3030	2780	4670	11850	5050	4865
9	3050	2890	4820	11010	5010	4910
10	2970	2940	4780	10015	5010	4980
11	2980	3000	4900	9875	5015	5000
12	2990	3020	5020	9005	5015	5010
13	3020	3120	5045	8880	5010	4950
14	3100	3180	4945	7990	5015	4900
15	2900	3210	4855	7345	5020	4845
16	3000	3270	4780	6920	5020	4800
17	3040	3455	4650	6745	5010	4785

Month	Kansas City	Chicago	Houston	Oklahoma City	Omaha	Little Rock
18	3060	3575	4535	6010	5015	4740
19	2950	3765	4475	5670	5040	4700
20	2970	3810	4330	5345	5010	4695
21	2990	3910	4325	5110	5020	4690
22	3060	3990	4155	4760	5010	4680
23	3000	4010	4090	4320	5015	4670
24	3010	4030	4010	4030	5010	4660
25	**2980**	**4285**	**3720**	**3005**	**5010**	**4590**
26	**2965**	**4420**	**3530**	**2515**	**5010**	**4570**
27	**2,945**	**4,560**	**3,330**	**2,030**	**5,005**	**4,555**

As a part of the descriptive analysis, summary statistics were generated from SAS (Table 8.3). The mean values provide some basis for a monthly demand rate, but at this point consideration of overall behavior within data distributions is required to more accurately capture relevant variation. To that end, other statistics can provide some picture of the distribution of the data. For example, the Kurtosis coefficient (see Chapter 5, "What Is Descriptive Analytics?") for Omaha's demand suggests a peaked distribution. This indicates that the variance about the mean is closely grouped toward the mean, implying a lack of variability in forecast values (a good thing). Note that the Standard Error statistic (see Chapter 5, Section 5.3) for Omaha is the smallest. Other statistics such as the Skewedness Coefficients suggest most of the distributions are negatively skewed. The median value peaks at a larger value than the mean and implies that the mean and mean-related statistics might not be as accurate in measuring the entire distribution's behavior as other measures (like the median).

Table 8.3 SAS Summary Statistics of Actual Monthly Customer Demand in Motors

Statistics	Kansas City	Chicago	Houston	Oklahoma City	Omaha	Little Rock
Mean	3015.41	3142.70833	4497.91666	9117.70833	5021.6666	4779.16666
Standard Error	12.3502	116.532393	70.8724849	710.692149	3.2646918	25.5087131
Median	3005	3070	4505	8942.5	5015	4745
Mode	2990	#N/A	4780	#N/A	5015	4680
Standard Dev.	60.5036	570.889807	347.202849	3481.66626	15.993658	124.966662

Statistics	Kansas City	Chicago	Houston	Oklahoma City	Omaha	Little Rock
Sample Var.	3660.68	325915.172	120549.818	12121999.9	255.79710	15616.6666
Kurtosis	0.09225	−1.002556	−1.279537	−1.510227	4.5845815	−0.9808275
Skewness	0.14712	0.11688577	−0.0192427	0.07550404	2.0952162	0.57001986
Range	240	1900	1100	9990	65	400
Minimum	2900	2130	3945	4030	5010	4610
Maximum	3140	4030	5045	14020	5075	5010
Sum	72370	75425	107950	218825	120520	114700
Count	24	24	24	24	24	24
Largest (1)	3140	4030	5045	14020	5075	5010
Smallest (1)	2900	2130	3945	4030	5010	4610
Confidence Level (95.0%)	25.5484	241.065623	146.610905	1470.17872	6.7535295	52.7687935

To better depict the general shape of the data and to understand its behavior, the analyst creates line graphs (see Chapter 5, Section 5.2) of the six customer demand files using SAS in Figures 8.1 to 8.6. As expected based on the summary statistics and now visually from the graphs, some of the customer demand functions look fairly linear, others are clearly nonlinear, and some possess so much variation they are unrecognizable. Identifying the almost perfect linear customer demand behavior in the warehouses in Chicago (Figure 8.2) and Oklahoma City (Figure 8.4) suggests the use of a simple linear regression model for forecasting purposes. The very clear, bell-shaped, nonlinear functions for Houston (Figure 8.3) and Little Rock (Figure 8.6) suggest that a nonlinear regression model should be determined by the BA team to find the best-fitting forecasting model. Finally, the excessively random customer demand behavior for Kansas City (Figure 8.1) and Omaha (Figure 8.5) suggests that considerable effort is needed to find a model that may or may not explain the variation in the data well enough for a reliable forecast. There appear to be many time series variations (see Appendix E, Section E.2) in customer demand for the warehouses in these two cities.

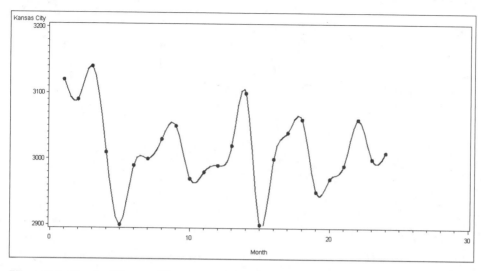

Figure 8.1 Graph of Kansas City customer demand

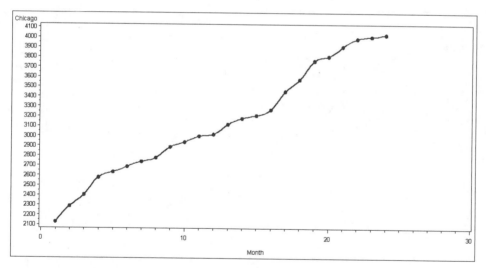

Figure 8.2 Graph of Chicago customer demand

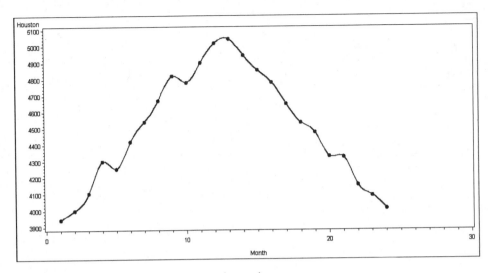

Figure 8.3 Graph of Houston customer demand

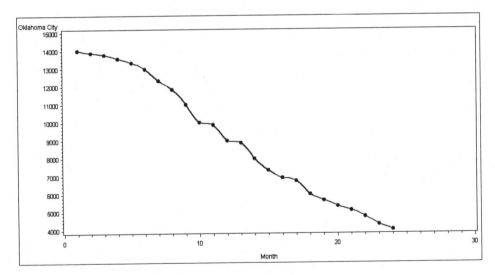

Figure 8.4 Graph of Oklahoma City customer demand

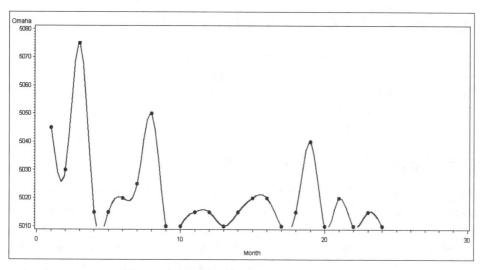

Figure 8.5 Graph of Omaha customer demand

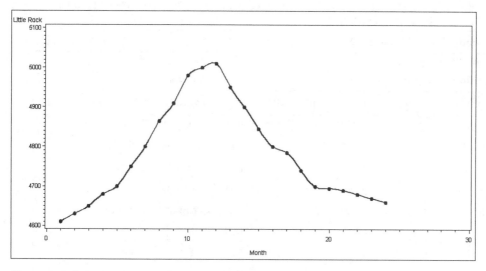

Figure 8.6 Graph of Little Rock customer demand

The fact that two of the four warehouse time series data files have more time series variations than the other four warehouse files does not prevent in this case (and in most others) a fairly accurate forecast. Because four of the six customer demand warehouses appear to have a fairly observable pattern of behavior, they will help improve the overall accuracy even with the substantial variations of the other two warehouses adding in some forecast error.

8.4 Predictive Analytics Analysis

In this section, we continue with our illustrative example. Here we use the predictive analytics analysis step that requires model development effort and then model validation for the example. To complete the predictive analytics analysis, we determine forecasts of warehouse demand.

8.4.1 Developing the Forecasting Models

The descriptive analytics analysis has suggested a course of action in identifying appropriate forecasting models in this next step of the BA process. To ensure the best possible forecasting models and confirm the descriptive analytics analysis results, the analyst will utilize curve fitting with SAS. Each of the six customer demand data files is analyzed through the SAS program to generate potential regression models, as presented in Tables 8.4 through 8.9. In this study, only seven regression models are used.

Table 8.4 SAS Curve-Fitting Analysis for Kansas City Motor Demand Forecasting Model: Model Summary and Parameter Estimates

Dependent Variable: KanCity									
Equation	Model Summary					Parameter Estimates			
	R-Square	F	df1	df2	Sig.	Constant	b1	b2	b3
Linear	.073	1.727	1	22	.202	3044.275	–2.309		
Logarithmic	.174	4.637	1	22	.043	3084.805	–30.398		
Inverse	.244	7.117	1	22	.014	2992.958	142.744		
Quadratic	.190	2.463	2	21	.109	3095.667	–14.168	.474	
Cubic	**.260**	**2.346**	**3**	**20**	**.103**	**3149.071**	**–37.459**	**2.757**	**–.061**
S	.239	6.905	1	22	.015	8.004	.047		
Growth	.070	1.658	1	22	.211	8.021	–.001		
The independent variable is Year.									

Table 8.5 SAS Curve-Fitting Analysis for Chicago Motor Demand Forecasting Model: Model Summary and Parameter Estimates

Dependent Variable: Chicago

Equation	Model Summary					Parameter Estimates			
	R-Square	F	df1	df2	Sig.	Constant	b1	b2	b3
Linear	.982	1230.679	1	22	.000	2142.409	80.024		
Logarithmic	.844	119.094	1	22	.000	1701.041	631.563		
Inverse	.480	20.270	1	22	.000	3439.516	−1886.509		
Quadratic	.984	647.977	2	21	.000	2199.355	66.883	.526	
Cubic	**.986**	**467.241**	**3**	**20**	**.000**	**2116.996**	**102.802**	**−2.994**	**.094**
S	.561	28.098	1	22	.000	8.140	−.660		
Growth	.979	1046.077	1	22	.000	7.714	.026		

The independent variable is Year.

Table 8.6 SAS Curve-Fitting Analysis for Houston Motor Demand Forecasting Model: Model Summary and Parameter Estimates

Dependent Variable: Houston

Equation	Model Summary					Parameter Estimates			
	R-Square	F	df1	df2	Sig.	Constant	b1	b2	b3
Linear	.003	.076	1	22	.786	4461.938	2.878		
Logarithmic	.127	3.191	1	22	.088	4158.249	148.801		
Inverse	.239	6.892	1	22	.015	4625.232	−809.218		
Quadratic	.929	138.408	2	21	.000	3632.994	194.173	−7.652	
Cubic	**.930**	**88.001**	**3**	**20**	**.000**	**3621.937**	**198.995**	**−8.124**	**.013**
S	.252	7.415	1	22	.012	8.438	−.186		
Growth	.004	.088	1	22	.769	8.400	.001		

The independent variable is Year.

Table 8.7 SAS Curve-Fitting Analysis for Oklahoma City Motor Demand Forecasting Model: Model Summary and Parameter Estimates

Dependent Variable: OKCity

Equation	Model Summary					Parameter Estimates			
	R-Square	F	df1	df2	Sig.	Constant	b1	b2	b3
Linear	.986	1567.704	1	22	.000	15229.746	–488.963		
Logarithmic	.826	104.691	1	22	.000	17817.142	–3811.033		
Inverse	.411	15.382	1	22	.001	7440.961	10657.408		
Quadratic	.987	775.938	2	21	.000	15420.511	–532.986	1.761	
Cubic	**.996**	**1697.830**	**3**	**20**	**.000**	**14294.516**	**–41.911**	**–46.359**	**1.283**
S	.338	11.256	1	22	.003	8.860	1.147		
Growth	.978	990.240	1	22	.000	9.763	–.058		

The independent variable is Year.

Table 8.8 SAS Curve-Fitting Analysis for Omaha Motor Demand Forecasting Model: Model Summary and Parameter Estimates

Dependent Variable: Omaha

Equation	Model Summary					Parameter Estimates			
	R-Square	F	df1	df2	Sig.	Constant	b1	b2	b3
Linear	.220	6.205	1	22	.021	5034.928	–1.061		
Logarithmic	.300	9.424	1	22	.006	5045.741	–10.546		
Inverse	.253	7.435	1	22	.012	5015.632	38.358		
Quadratic	.299	4.477	2	21	.024	5046.077	–3.634	.103	
Cubic	**.340**	**3.432**	**3**	**20**	**.037**	**5056.845**	**–8.330**	**.563**	**–.012**
S	.253	7.454	1	22	.012	8.520	.008		
Growth	.220	6.212	1	22	.021	8.524	.000		

The independent variable is Year.

Table 8.9 SAS Curve-Fitting Analysis for Little Rock Motor Demand Forecasting Model: Model Summary and Parameter Estimates

Dependent Variable: LittleRock

Equation	Model Summary					Parameter Estimates			
	R-Square	F	df1	df2	Sig.	Constant	b1	b2	b3
Linear	.001	.019	1	22	.893	4785.580	−.513		
Logarithmic	.068	1.602	1	22	.219	4689.689	39.198		
Inverse	.167	4.399	1	22	.048	4817.464	−243.419		
Quadratic	.754	32.130	2	21	.000	4516.566	61.567	−2.483	
Cubic	**.801**	**26.824**	**3**	**20**	**.000**	**4426.153**	**100.999**	**−6.347**	**.103**
S	.171	4.539	1	22	.045	8.480	−.051		
Growth	.001	.015	1	22	.903	8.473	−9.677E-005		

The independent variable is Year.

When the analyst reviews the R-Square values for each of the potential curve-fitting models, it turns out that the cubic model is the best fitting for all six data files. It is not surprising that in the cases of Houston and Little Rock, where the descriptive analytics graphs clearly show typical cubic (or quadratic) function behavior, the only significant (F-ratio, p<.000) models were cubic or quadratic (see Chapter 6, Section 6.4.2). In other cases (Chicago and Oklahoma City), it is surprising that a nonlinear cubic model does a slightly better job than the descriptive analytics step linear model. On the other hand, note that for both locations, the linear model, according to the R-Square statistics, is either the next best choice or the next to the next best choice. Indeed, in both cases, the F-ratio clearly shows that the resulting linear model can provide a statistically significant forecasting capability. Other models also have significant (p<.000) F-ratios, suggesting the possibility of accurate forecasting. Because the objective of this case study is to develop a procedure that analysts could use each month to develop an optimal supply chain schedule of shipments and to accurately forecast customer demand, the BA analyst can use the highest R-Square statistic as a means to determine the most accurate forecasting model from those fitted with the data.

In this case study, the resulting cubic regression models estimated by the SAS program based on the parameters from the curve-fitting effort are presented in Table 8.10.

Table 8.10 Resulting Cubic Forecasting Models from SAS Curve-Fitting Analysis

Location	Constant	b1	b2	b3	R-Square
Kansas City	3149.071	–37.459	2.757	–.061	.260
Chicago	2116.996	102.802	–2.994	.094	.986
Houston	3621.937	198.995	–8.124	.013	.930
Oklahoma City	14294.516	–41.911	–46.359	1.283	.996
Omaha	5056.845	–8.330	.563	–.012	.340
Little Rock	4426.153	100.999	–6.347	.103	.801

The generalized formula for a cubic regression model follows:

$$Y_p = a + b_1 X + b_2 X^2 + b_3 X^3$$

When the analyst inserts the curve-fitted parameters for Little Rock, the resulting cubic regression model for forecasting warehouse customer demand is this:

$$Y_p = 4426.153 + 100.999 X - 6.347 X^2 + 0.801 X^3$$

where:

X = month number in the form of the time series data file (25, 26, and 27)

8.4.2 Validating the Forecasting Models

One of the fundamental requirements of a BA analysis is to show or prove the possibility of improving business performance (see Chapter 1, Section 1.1). One criterion for improving forecasting is to improve forecasting accuracy. To compare the current forecasting method with the newly devised one, the analyst uses each cubic model to forecast the respective location of customer demand. Substituting the numbered time values (25, 26, and 27) for X in each cubic model, the analyst is able to compute the three forecast values. These forecasts are then compared with the actual values in Table 8.2. The resulting comparison is expressed in the MAD statistics (see Appendix E, Section E.8), as presented in Table 8.11.

Table 8.11 Resulting Cubic Model Forecasts and MAD Statistics (Rounded Up to Next Integer Value)

Month	Kansas City	Chicago	Houston	Oklahoma City	Omaha	Little Rock
25	2983	4285	3723	4320	5013	4594
26	2967	4419	3533	4417	5010	4572
27	2947	4561	3329	4621	5007	4554
MAD	2.33	0.66	2.33	1936	1.66	2.33

The MAD statistics for all warehouse facilities except Oklahoma City are extremely small, suggesting the cubic models for these locations are very accurate. The Oklahoma City MAD statistic, on the other hand, is so great relative to the other MADs that it suggests further analysis is needed to find a better forecasting model for Oklahoma City.

To explore this forecast exception in Oklahoma City, the analyst examines the next two best models (based on R-Square) from the SAS curve-fitting effort in Table 8.7. These two include the linear regression model (R-Square 0.986):

$$Y_p = 15229.746 - 488.963X$$

and the quadratic regression model (R-Square 0.987):

$$Y_p = 15420.511 - 532.986 X + 1.761 X^2$$

In Table 8.12, the resulting warehouse forecasts of customer demand for each of the three years are presented along with their MAD statistics for both the linear and the quadratic models. Clearly, the linear regression model's small MAD suggests that it is the better model for forecasting than either the quadratic or the cubic models. This result is not surprising, given the prior descriptive analytics step, which appeared to suggest that a linear model would be the best type of forecasting model.

Table 8.12 Resulting Linear and Quadratic Forecasts with MAD Statistics for Oklahoma City (Rounded Up to the Next Integer Value)

Year (X)			
Model	**25**	**26**	**27**
Linear Forecast	3006	2517	2028
	MAD = 2.33		
Quadratic Forecast	3197	2754	2314
	MAD = 238.33		

Having found the models that provide low error rates, the analyst now needs to validate them by demonstrating they can improve forecasting accuracy and, therefore, enhance business performance by minimizing costly shipping efforts.

To validate the forecasting accuracy and demonstrate forecasting improvement of the cubic and linear models, the analyst undertakes a comparison with the currently used smoothing average method. Utilizing a similar smoothing average formula to that mentioned in Section E.6 in Appendix E, the analyst can compute the forecast values for warehouse customer demand using the following simple formula:

$$\bar{Y}_t = (Y_{t-1} + Y_{t-2} + Y_{t-3})/3$$

where:

$\overline{Y}_t$ = the forecast value in time period t

Y_{t-1} = the actual value in the time period just prior to time period t

Y_{t-2} = the actual value of two time periods prior to time period t

Y_{t-3} = the actual value of three time periods prior to time period t

Using the formula, the resulting smooth average forecast values are presented in Table 8.13 along with their respective MAD statistics.

Table 8.13 Resulting Smooth Average Forecasts and MAD Statistics (Rounded Up to Next Integer Value)

Month	Kansas City	Chicago	Houston	Oklahoma City	Omaha	Little Rock
25	3024	4010	4085	4370	5012	4670
26	2997	4109	3940	3785	5012	4640
27	2985	4245	3754	3184	5010	4607
MAD	21.33	111.33	158.66	590.33	1.66	31

These smoothing average forecasts and their MADs can be compared with the forecasts and MADs for the cubic and linear models. Comparing the MADs in Table 8.13 with the MADs in Tables 8.11 and 8.12, the analyst can make several points about forecasting improvement. The cubic regression models are the lowest for Kansas City, Chicago, Houston, and Little Rock; therefore, they have more accurate forecasting results. For those four locations, the cubic model is recommended. In the case of Oklahoma City, the linear regression model results in the lowest MAD value, reflecting improved forecasting accuracy over the other models. Finally, the MADs for both the cubic regression (Table 8.11) and the smoothing average methods (Table 8.13) result in the same MAD value (1.66) for Omaha, which suggests either method is accurate in forecasting this location's customer demand. Because either method can be used, the manufacturer's BA analyst selected to employ the cubic regression model for forecasting Omaha's warehouse customer demand.

8.4.3 Resulting Warehouse Customer Demand Forecasts

The selected forecasting models and their forecast values for the future 28th month (X = 28 in the models) are presented in Table 8.14. The direction of movement from the 27th to the forecast of the 28th month appears to be downward for most of the warehouse locations. The resulting forecast values for the six locations are generally consistent with graphs from the descriptive analytics analysis, although some of the time series variation behavior (for example, Kansas City) can hardly be predicted up or down in movement.

Table 8.14 Resulting Forecasts for the Future 28th Month (Rounded Up to Next Integer Value)

Location	Forecast Model	Forecast (X = 28)	Direction
Kansas City	$Y_p = 3149.071 - 37.459X + 2.757 X^2 - .061 X^3$	2923	Down
Chicago	$Y_p = 2116.996 + 102.802X - 2.994 X^2 + .094 X^3$	4712	Up
Houston	$Y_p = 3621.937 + 198.995X - 8.124 X^2 + .013 X^3$	3110	Down
Oklahoma City	$Y_p = 15229.746 - 488.963X$	1539	Down
Omaha	$Y_p = 5056.845 - 8.330X + .563 X^2 - .012 X^3$	5002	Down
Little Rock	$Y_p = 4426.153 + 100.999X - 6.347 X^2 + .103 X^3$	4540	Down
Total Forecast Customer Demand for the 28th Month		21826	

8.5 Prescriptive Analytics Analysis

Based on the predictive analytics analysis, the total forecast demand (see Table 8.14) of 21,826 motors for all six warehouse locations has to be balanced out by product capacity of the two production facilities. The BA team decided that the St. Louis production center would produce 10,000 motors for the 28th month, and the Dallas production center would produce the remaining 11,826 motors.

8.5.1 Selecting and Developing an Optimization Shipping Model

In terms of data, the analyst now possesses the supply, forecast demand, and cost information on which to begin selecting a modeling approach to achieve an optimal shipping schedule. Reviewing the requirements of the problem setting at this point in the analysis, the BA team is looking at a multivariable (number of motors to ship from two supply sources to three demand destinations), multidimensional (scheduling motor shipments from two supply sources to six demand markets or supply and demand), constrained (the exact number of motors required is deterministic at this point), integer (shipping whole motors, not motor parts), and optimal solution (seeking to minimize cost of shipping). The ideal BA methodology to satisfy these require-ments is *integer programming* (IP). (See Appendix D, "Integer Programming.")

To help you conceptualize a two-dimensional problem, we present a *transporta-tion method* (an operations research methodology) table that combines location cost per unit shipped and supply and demand information. This case study problem is shown in Table 8.15. The decision variables for the model are also added. So, for example, X_{11} represents the number of manufacturer motors to ship from the St. Louis production center to meet the forecast customer demand at the Kansas City warehouse. In this table, the sum of the motors produced in St. Louis and shipped to any of the six customer demand locations must add up to 10,000. Likewise, for Dal-las, the shipments must equal 11,826. Also, for each column, the sum of the motors shipped must equal the forecast demand in that column. For example, the sum of X_{11} and X_{21} for the Kansas City warehouse must equal the forecast demand of 2,923 motors.

Table 8.15 Transportation Method Table for Conceptualization of Supply Chain Shipping Problem

	Kansas City, Missouri	Chicago, Illinois	Houston, Texas	Oklahoma City, Oklahoma	Omaha, Nebraska	Little Rock, Arkansas	Production Center Supply
St. Louis, Missouri	$4\,X_{11}$	$6\,X_{12}$	$9\,X_{13}$	$8\,X_{14}$	$5\,X_{15}$	$6\,X_{16}$	10000
Dallas, Texas	$5\,X_{21}$	$8\,X_{22}$	$2\,X_{23}$	$5\,X_{24}$	$8\,X_{25}$	$5\,X_{26}$	11826
Forecast of Customer Demand	2923	4712	3110	1539	5002	4540	21826

The IP model can be developed with the transportation method table as a frame-work. In this type of shipping problem, there are two supply-side constraints and six demand-side constraints required to ensure the supply is allocated to meet the demand. The same formulation procedure for LP models in Appendix B, "Linear Pro-gramming," and for integer programming in Appendix D is applied here to generate the following integer model:

$$\text{Minimize: } Z = 4X_{11} + 6X_{12} + 9X_{13} + 8X_{14} + 5X_{15} + 6X_{16}$$
$$+ 5X_{21} + 8X_{22} + 2X_{23} + 5X_{24} + 8X_{25} + 5X_{26}$$

subject to:

$$X_{11} + X_{12} + X_{13} + X_{14} + X_{15} + X_{16} = 10000 \text{ (St. Louis supply requirement)}$$

$$X_{21} + X_{22} + X_{23} + X_{24} + X_{25} + X_{26} = 11826 \text{ (Dallas supply requirement)}$$

$$X_{11} + X_{21} = 2923 \text{ (Kansas City demand requirement)}$$

$$X_{12} + X_{22} = 4712 \text{ (Chicago demand requirement)}$$

$$X_{13} + X_{23} = 3110 \text{ (Houston demand requirement)}$$

$$X_{14} + X_{24} = 1539 \text{ (Oklahoma City demand requirement)}$$

$$X_{15} + X_{25} = 5002 \text{ (Omaha demand requirement)}$$

$$X_{16} + X_{26} = 4540 \text{ (Little Rock demand requirement)}$$

and $X_{11}, X_{12}, X_{13}, X_{14}, X_{15}, X_{16}, X_{21}, X_{22}, X_{23}, X_{24}, X_{25}, X_{26} \geq 0$ and all integer

8.5.2 Determining the Optimal Shipping Schedule

To run this model, we utilize LINGO (see Appendix B, Section B.5.3) software. As it turns out, in this situation the unique formulation of the transportation method model mathematically forces an all-integer solution without the need for using the IP software algorithm. This permits the regular LP software to be used to solve this problem, although LINGO has both IP and LP solution software. The LINGO LP model input is presented in Figure 8.7, and the results are presented in Figure 8.8.

```
LINGO Model - LINGO1
Min = 4 * x11 + 6 * x12 + 9 * x13 + 8 * x14 + 5 * x15 + 6 * x16 + 5 * x21 + 8 * x22 + 2 * x23 + 5 * x24 + 8 * x25 + 5 * x26;
x11 + x12 + x13 + x14 + x15 + x16 = 10000;
x21 + x22 + x23 + x24 + x25 + x26 = 11826;
x11 + x21 = 2923;
x12 + x22 = 4712;
x13 + x23 = 3110;
x14 + x24 = 1539;
x15 + x25 = 5002;
x16 + x26 = 4540;
```

Figure 8.7 LINGO input for supply chain shipping model problem

```
Solution Report - LINGO1                                   □ □ ☒
   Global optimal solution found.
   Objective value:                          104226.0
   Total solver iterations:                         1


                  Variable          Value        Reduced Cost
                       X11       286.0000            0.000000
                       X12       4712.000            0.000000
                       X13       0.000000            8.000000
                       X14       0.000000            4.000000
                       X15       5002.000            0.000000
                       X16       0.000000            2.000000
                       X21       2637.000            0.000000
                       X22       0.000000            1.000000
                       X23       3110.000            0.000000
                       X24       1539.000            0.000000
                       X25       0.000000            2.000000
                       X26       4540.000            0.000000

                       Row  Slack or Surplus       Dual Price
                         1       104226.0           -1.000000
                         2       0.000000            0.000000
                         3       0.000000           -1.000000
                         4       0.000000           -4.000000
                         5       0.000000           -6.000000
                         6       0.000000           -1.000000
                         7       0.000000           -4.000000
                         8       0.000000           -5.000000
                         9       0.000000           -4.000000
```

Figure 8.8 LINGO output for supply chain shipping model problem

Extracting the shipping schedule for the supply chain problem, we present the number of motors to be shipped from the two supply source locations to the seven demand destinations in Table 8.16 (the bold numbers in the rows and columns in the table). For example, to achieve a cost-minimized shipping schedule, the manufacturer has to ship 286 motors from St. Louis to the Kansas City warehouse in the 28th month. Likewise, all the other eight scheduled shipments in Table 8.16 must be shipped exactly as scheduled to ensure the optimization of the total costs. Note in Table 8.16 that the allocation of motors exactly adds up to the last column supply values and the bottom row demand values.

Table 8.16 Shipping Schedule for 28th Month Supply Chain Shipping Problem

	Kansas City, Missouri	Chicago, Illinois	Houston, Texas	Oklahoma City, Oklahoma	Omaha, Nebraska	Little Rock, Arkansas	Production Center Supply
St. Louis, Missouri	286	4712			5002		10,000
Dallas, Texas	2637		3110	1539		4540	11,826
Forecast of Customer Demand	2923	4712	3110	1539	5002	4540	21,826

Also, the value of 104,266 in Figure 8.8 is the total optimized cost for this shipping schedule (taking the units shipped in each cell of Table 8.16 and multiplying them by the number of units in those cells). The resulting shipping schedule for the supply chain problem in month 28 is detailed in Table 8.17.

Table 8.17 Resulting Shipping Schedule for Month 28

Ship From	To	Units × Per Unit Cost = Total Cost
St. Louis	Kansas City	286 × $4 = $1,144
St. Louis	Chicago	4,712 × $6 = $28,272
St. Louis	Omaha	5,002 × $5 = $25,010
Dallas	Kansas City	2,637 × $5 = $13,185
Dallas	Houston	3,110 × $2 = $6,220
Dallas	Oklahoma City	1,539 × $5 = $7,695
Dallas	Little Rock	4,540 × $5 = $22,700
Total Supply Chain Shipping Cost = $104,226		

8.5.3 *Summary of BA Procedure for the Manufacturer*

The intent of this BA application is to develop a BA procedure for the manufacturer to utilize every month in planning the supply chain problem of setting up an optimal shipping schedule in the supply chain network. This BA procedure based on the BA team analysis presented here involves both data collection efforts: statistical analysis and the application of optimization software. Specifically:

1. Collect shipping cost information from the firm's cost accounting department.
2. Collect and update monthly actual demand values from warehouse customers.
3. Collect supply center supply capacity to ensure sufficient supply capacity exists to handle monthly demand.
4. Rerun curve-fitting software on new and old actual demand data to determine the best forecasting model based on the R-Square and other statistics as needed.
5. Forecast warehouse customer demand and affirm through analysis that the resulting estimates are truly based on the best forecasting model. Revise as needed to select the best model.
6. Incorporate the cost, supply, and forecast demand information into a linear programming model similar to what was developed in Section 8.5.1.
7. Run the IP or LP model and extract the shipping schedule from the model output.

8.5.4 Demonstrating Business Performance Improvement

A BA is not complete without showing that business performance can or will be improved. This case study has a basis for comparison. In comparing the MAD statistics from the present forecasting procedure and the BA proposed procedure, we can observe a potential for improvement in shipping. (The MAD statistic is explained in Section E.8 in Appendix E.) In Table 8.18, the MAD values based on the three months (months 25, 26, and 27) used for model validation are presented. The MAD statistics (see Appendix E, Section E.8) represent the average monthly overage or underage of motors that could have been avoided if the BA proposed procedure would have been in place. Such needless shipments waste effort and add to cost inefficiencies for the manufacturer. Establishing a procedure that lowers the MAD statistics would represent an opportunity for improving business performance.

As can be seen in Table 8.18, the current procedure of using a smoothed averaging to generate a forecast results in fairly large MAD statistics compared to the proposed BA procedure. The total of the MADs in Table 8.18 clearly shows a significant reduction in monthly overages or underages when using the proposed BA procedure in forecasting motor customer demand. Reducing the inaccuracy in forecasting also translates into minimizing wasted costs of shipping the motors that are either not needed in the warehouses during low customer demand periods or rush-ordered when shortages occur. These results reveal that the implementation of the proposed BA procedure for the supply chain shipping schedule problem could have improved business performance for the manufacturer over what was previously used to forecast the last three months.

Table 8.18 Comparisons of MAD Statistics Between Present and BA Proposed Forecasting Procedures

MADs	Kansas City, Missouri	Chicago, Illinois	Houston, Texas	Oklahoma City, Oklahoma	Omaha, Nebraska	Little Rock, Arkansas	Total
MADs for present smoothing model (Table 8.14)	21.33	111.33	158.66	590.33	1.66	31	914.31
MADs for proposed cubic and linear models (Tables 8.12 and 8.13)	2.33	0.66	2.33	2.33	1.66	2.33	11.64

As a final recommendation from the BA team on the prescriptive analytics step, the analyst or BA team responsible for utilizing the new BA procedure should continuously run updates to check and confirm the benefits of using the BA procedure on a monthly basis. Continually showing the worth of BA is recommended for the success of BA in firms (see Table 2.2 in Chapter 2, "Why Is Business Analytics Important?").

Summary

This chapter presented a case study illustrating the use of BA to solve a supply chain shipping problem. The case study utilized the three-step BA process to develop a BA procedure that could be repeated monthly to improve a manufacturer's business performance.

The particular use of methodologies in this case study could have been different and could have highlighted the fact that BA is meant as a step-wise guide in the application of statistical, information system, and management science methodologies. Like a walk in a forest, there may be many paths, but the goal is to reach the other side using knowledge and information (from a BA analysis) to support your steps.

This chapter ends the text material of this book, but the appendixes offer readers a rich foundation of methodologies useful in BA. Some have been demonstrated in the text material, and others have not, but all can be useful for differing analyses. The more methodologies that BA analysts know, the more likely they are to utilize the right one in the right situation. The appendixes are a starting point on which to build a foundation of methodological tools to strengthen and continually augment BA knowledge.

Discussion Questions

1. Some of the graphs (for example, Chicago going up) in the descriptive analytics analysis tended to show fairly linear behavior, yet a cubic model, rather than a linear model, was used in most cases. Did it make sense to use the cubic model instead of the linear model? Why?

2. The SAS curve-fitting process in this study involved the development of seven different regression models. Should all the regression models have been tested with the validation data in months 25, 26, and 27?

3. What other methodologies in this book might have been applied to analyze this case study problem?

4. Why is it necessary to show and continuously demonstrate business improvement in the analysis when it is clear that forecasting results are improved?

Problems

1. How was the Kansas City cubic regression forecast of 2,983 for month 25 computed? Show the formula with input values.

2. How was the Chicago cubic regression forecast of 4,561 for month 27 computed? Show the formula with input values.

3. How was the Oklahoma City quadratic regression forecast of 2,754 for month 26 computed? Show the formula with input values.

4. How was the MAD statistic for the Oklahoma City linear regression model computed on the three-month validation data? Show the formula with input values.

Part IV

Appendixes

A

Statistical Tools

A.1 Introduction

The purpose of this appendix is to provide a brief overview of some basic statistical tools, including counting, probability concepts, probability distributions, and statistical testing. Other statistical methods and testing content will be presented in chapters and other appendixes.

A.2 Counting

Counting is an important prerequisite for computing probabilities. A probability example is usually made up of the ratio of a few observed behaviors over the total of all possible behaviors:

$$\text{Probability Estimate} = \frac{\text{A Few Observed Behaviors (Numerator)}}{\text{Total of All Possible Behaviors (Denominator)}}$$

To use this probability formula, we must determine both the numerator and the denominator of the probability estimate ratio. The three methods of counting that can be used to determine probability values include permutations, combinations, and repetitions.

A.2.1 Permutations

A *permutation* is a specific ordering of elements from a collection of elements. One usually wants to determine or count the total permutations that are possible from a given collection of elements. A collection of elements, for example, might be the items that make up a data set. *Permutations* can be defined as the number of ways r elements at a time are taken from a collection of n elements, such that (1) the ordering of the elements is important; (2) there can be no repetitions of the same element in a set of r; and (3) the value of r will either equal n or will be less than n. To determine the total number of permutations possible from a collection of n elements taken r at a time, note the following permutation formula:

$$_nP_r = \frac{n!}{(n-r)!}$$

where:

! = a factorial sign. (A *factorial* is a mathematical abbreviation indicating that the value it is attached to should be multiplied by all the values that precede it. For example, 4! is an abbreviation for $4 \times 3 \times 2 \times 1$, or 24.)

P = the possible number of permutations of elements out of a collection of elements

n = the total number of elements in a collection

r = the number of elements taken from the collection at one time

In using this formula, it is important to remember that the three characteristics included in the permutation formula (order is important, objects should not be repeated, and $r \le n$) are strictly observed. These characteristics of the permutation formula will change for each of the other counting formulas discussed.

Question: What are the permutations of five letters A, B, C, D, and E taken two at a time?

Answer: You can use the permutations formula to calculate this:

$$_nP_r = \frac{n!}{(n-r)!} = \frac{5!}{(5-2)!} = \frac{5 \times 4 \times 3 \times 2 \times 1}{3!} = \frac{120}{6} = 20 \text{ permutations}$$

The permutations can be listed using this systematic procedure. Begin with the listing of the permutations with the first letters first. These permutations are AB, AC, AD, AE, BC, BD, BE, CD, CE, and DE. Then reverse the letters to obtain the remaining permutations, which are BA, CA, DA, EA, CB, DB, EB, DC, EC, and ED.

A.2.2 Combinations

A combination, like a permutation, is a specific ordering of elements from a collection of elements. *Combinations* can be defined as the number of ways r elements at a time are taken from a collection of n elements, such that (1) the ordering of the elements is not important, (2) there can be no repetitions of the same element in a set of r, and (3) the value of r will either equal n or will be less than n. To determine the total number of combinations that are possible from a collection of n elements, taken r at a time, note the following formula:

$$_nC_r = \frac{n!}{r!(n-r)!}$$

where:

C = the possible number of combinations of elements out of a collection of elements

n = the total number of elements in a collection

r = the number of elements taken from the collection at one time

In using this formula, it is important to remember that the characteristic that differentiates combinations from permutations is that the order of elements is not important. The effect of ordering as not being important will mean that the possible number of combinations will always be less than the possible number of permutations.

Question: What are the combinations of five letters A, B, C, D, and E taken two at a time?

Answer: Using the permutations formula, we find the following to be true:

$$_nC_r = \frac{n!}{r!(n-r)!} = \frac{5!}{2!(5-2)!} = \frac{5 \times 4 \times 3 \times 2 \times 1}{2 \times 1(3 \times 2 \times 1)} = \frac{120}{12} = 10 \text{ permutations}$$

The combinations can be listed using the same systematic procedure previously discussed. List the combinations with the first letters first. The ten combinations for this problem are AB, AC, AD, AE, BC, BD, BE, CD, CE, and DE.

A.2.3 Repetitions

When repetitions of the same element in a collection of elements are possible, a different type of counting formula is necessary. A *repetition*, like a permutation or a combination, is a specific ordering of elements from a collection of elements. *Repetitions* can be defined as the number of ways r elements at a time are taken from a collection of n elements such that (1) the ordering of the elements is important; (2) there can be as many as r repetitions of the same element in a set of r; and (3) the value of r will either equal n or will be less than n. To determine the total number of repetitions possible from a collection of n elements, taken r at a time, we have the following formula:

$$_nR_r = n^r$$

where:

R = the possible number of repetitions of r elements out of a collection of n elements

n = the total number of elements in a collection

r = the number of elements taken from the collection at one time

In using this formula, it is important to remember that the characteristic that differentiates repetitions from the other two counting procedures is that repetitions of the same elements are permitted.

Question: How many different repetitions of five food products (hamburgers, French fries, hot dogs, regular drink, and large drink) can be structured into a "food-deal" package if it takes three food products to make a package of goods? Assume that the same product can be used repeatedly in a single package. (For example, three hamburgers would be considered a "food-deal" package.)

Answer: To find the number of repetitions, use this formula:

$$_nR_r = n^r = 5^3 = 125 \text{ different "food-deal" packages}$$

So a total of 125 different "food-deal" packages can be structured to sell to customers from only five different products taking three products at a time.

A.3 Probability Concepts

Understanding probability usage in business analytics requires some fundamental knowledge of probability concepts. In this section we introduce approaches used to assess probabilities and then follow the discussion up with basic rules of addition and multiplication to help understand their manipulation.

A.3.1 Approaches to Probability Assessment

Before reviewing the basic rules of probability, be aware of some of the approaches used to assess probabilities. There are two general approaches to assessing probabilities: objective and subjective. The *objective approach to probability* is based on objective methods of collecting experimental data on trials and their outcomes, tabulating the data into frequency distributions and then into probabilities. The two theories that most commonly represent the objective approach to probability are Frequency Theory and the Principle of Insufficient Reason.

Frequency Theory basically states that, through a large number of trials, the relative frequency outcome of an event, A, can be used to determine the probability of A, represented here as P(A). This probability is based on experiential observations, and it is assumed that the experiment accurately represents the possible behaviors that are being observed. The probabilities that are determined using this approach are based on past observations that are converted into probabilities to be used in the future. Because the probabilities are based on past events, we sometimes refer to these probabilities as being *posteriori probabilities*.

Question: Using Frequency Theory, what is the probability that a food server will be tipped an amount that will fall in the tip class interval of $8 to $9.99 based on the following collection of data from a food server's experiences at the restaurant?

Number of Customers

Tips	Who Tipped
$6 to $7.99	50
$8 to $9.99	120
$10 to $12.99	30
Total	200

Answer: The relative frequency of the class interval of \$8–\$9.99 is 120 out of 200. If we assume the 200 tip experiments recorded are sufficient to accurately describe all tip behavior of the customers for the specific server, Frequency Theory holds that the probability of a tip between \$8 and \$9.99 being given to that server is 0.60, or 60 percent (120 / 200 = 0.60).

The *Principle of Insufficient Reason* states that each possible outcome in an experiment is equally probable if there is no evidence to challenge the assumption. Alternatively, if there is no reason to prefer one outcome over another, each outcome is equally likely or has the same probability. To determine probabilities using this principle, we must logically abstract the decimal value probabilities by dividing the frequency of occurrence by the total number of possible outcomes. Because there is no evidence to support that any of the possible outcomes is any more probable to occur than the rest, each will be given an equal probability.

Question: A stock broker must select one stock to invest in out of a sample of four stocks. The stock broker is unfamiliar with the four sample stocks. What is the probability that the stock broker will pick the best stock out of the four?

Answer: Assuming that there is only one "best" stock in the sample of four, and because there is one chance out of four to obtain it, the probability of choosing the best one out of four is 0.25 (1 / 4 = 0.25). In this problem, each of the four is equally likely to be the best stock, because no additional information on the stocks is available.

The probabilities that are determined using this principle are based on abstract reasoning of equality of outcomes and not on experience with observing outcomes of trials. The fact that the probabilities are based on prior logic and reasoning leads one to refer to these probabilities as *a priori probabilities*.

In the *subjective approach to probability,* estimation of probabilities is based on personal opinion or judgment. Under this approach, we assume that an individual's subjective judgment may be as accurate as or better than any other objective approach. BA analysts, whose subjective expertise is combined with the objective frequency information provided by a database, can combine these sources to greatly improve probability estimates for decision-making. To combine the probabilities, one must first learn some rules of how probabilities can be combined.

A.3.2 *Rules of Addition*

To apply one of the rules of addition, we must have a probabilistic situation that is characterized by a collectively exhaustive and mutually exclusive set of events. A *collectively exhaustive set of events* can be defined as the set of all possible events that can occur in an experiment. In a product failure test, either the product works or it fails to work. There could be two events (product works or product fails to work) that are collectively exhaustive, because each outcome of an event must fall into one of these two categories. *Mutually exclusive events* are events that are not related to one another. In other words, the probability of one event does not affect, alter, or impact in any way the probabilities of any other event. An event that is mutually exclusive cannot be counted or measured in more than one category during an experiment.

This concept is illustrated in Figure A.1. The Venn diagram on the left shows that P(A) and P(B) are mutually exclusive probabilities.

The *rule of addition* of probabilities is that for a mutually exclusive and collectively exhaustive set of events, the probability of any collection of possible events is found by the summation of the probabilities of those events. In other words, to find the probability of two events, we need only add respective individual probabilities. We can express this rule of addition as follows:

Rule of Addition (for mutually exclusive events A and B): P(A or B) = P(A) + P(B)

In reading statistical literature, the expression of two probabilities being added together can be represented by any of the following: P(A) + P(B), P(A or B) (referred to as the probability of A or B), or P(A $\cup$ B) (referred to as the probability of A "union" B, where the $\cup$ represents the union of both the probabilities of A and B).

Question: If there are four mutually exclusive and collectively exhaustive events of P(A) = 0.20, P(B) = 0.40, P(C) = 0.15, and P(D) = 0.25, what are the following probabilities: P(A) + P(D), P(B $\cup$ C), P(A or B or C), and P(A +~A)?

Answer: P(A) + P(D) = 0.20 + 0.25 = 0.45, P(B $\cup$ C) = 0.40 + 0.15 = 0.55, P(A or B or C) = 0.20 + 0.40 + 0.15 = 0.75, and P(A + ~A) = P(A) + (1 − P(A)) = 0.20 + 0.80 = 1.00

Sometimes events are not mutually exclusive. In the Venn diagram on the right in Figure A.1, the notation where two events are possible, P(A and B), is used. This is called *a joint probability* or *compound event*. The Venn diagram in Figure A.1 shows how the sample points of events A and B overlap. This overlapping of events means they are *not mutually exclusive*. The P(A) is jointly related to P(B). Note in the Venn diagram that if the total probabilities of P(A) and P(B) are added, the joint probability of P(A and B) would be double counted. To avoid this double counting for events that are not mutually exclusive, a revision of the rule of addition is needed, as stated earlier.

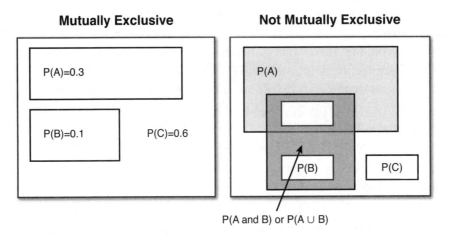

Figure A.1 Venn diagrams

The *rule of addition* of probabilities for events that are not mutually exclusive is that the probability of any collection of possible events is found by the summation of the probabilities of those events minus the joint probability of those events. In other words, to find the probability of two events, we need to add their respective individual probabilities together and subtract their joint probabilities. We can express this rule of addition as follows:

Rule of Addition (for events A and B that are not mutually exclusive):

P(A or B) = P(A) + P(B) − P(A and B)

We can expand the rule of addition for additional events by adding the probability of the new event and subtracting all joint probabilities.

A.3.3 Rules of Multiplication

To apply one of the rules of multiplication, we must have a probabilistic situation that is characterized by independent event outcomes. *Independent event outcomes* are outcomes in which the probability of one event does not affect the probability of another event. Unlike the rules of addition, the rules of multiplication apply only to multitrial experiments. In a multitrial experiment, we have several event outcomes. The *rule of multiplication* of probabilities is that, for independent event outcomes, the probability of any collection of possible events is found by the product of the probabilities of those events. In other words, to find the probability of two events, their respective individual probabilities should be multiplied together. This rule of addition is expressed as follows:

Rule of Multiplication (for independent events A and B):

P(A and B) = P(A) × P(B)

To illustrate this rule, assume that a purchasing manager faces a problem of selecting one of three component parts—A, B, or C—each week for the next two weeks. The probability of selecting Part A (P(A)) is 0.33. What is the probability that Part A will be selected in both Weeks 1 and 2? In this experiment, there are two events whose outcomes are independent, because the probability of Part A being selected in Week 1 will not alter the probability of Part A being selected in Week 2. The probability of Part A being selected in both weeks is P(A) × P(A), or 0.11 (0.33 × 0.33 = 0.11).

In reading statistical literature, the expression of two probabilities being multiplied together can be represented by any of the following: P(A) × P(B), P(A and B) referred to as the probability of A and B, or P(A ∩ B) (referred to as the probability of A and B, where the ∩ represents "and").

Question: If there are three independent probabilities of P(A) = 0.5, P(B) = 0.3, P(C) = 0.2, what are the values of P(A) × P(B), P(B and C), P(A ∩ C), and P(A and B and C)?

Answer: P(A) × P(B) = 0.5 × 0.3 = 0.15, P(B and C) = 0.3 × 0.2 = 0.06, P(A ∩ C) = 0.5 × 0.2 = 0.1, and P(A and B and C) = 0.5 × 0.3 × 0.2 = 0.03.

In an experiment with multiple events or trials, the outcomes of those events can be either independent or dependent. *Dependent event outcomes* occur when the probable outcome of one event affects the probable outcome of another event. For example, suppose there is a sample of five members of political Party A and five members of political Party B. Assume that the members of each party will vote for their own respective parties. Using this sample, an experiment is structured with two events. Event 1 randomly selects one of the ten members to ask whom that person will vote for, and then this party member is excluded from the sample. The process of excluding an element from a sample is called *sampling without replacement*, because one does not place the element, or party member in this example, back into the sample after its use. Event 2 in the experiment randomly selects one of the nine remaining members to ask whom that person will vote for. The probability of selecting a member of either Party A or Party B in Event 1 is 0.50 (5 / 10 = 0.50). The probability of selecting a member of either party in Event 2 is conditionally dependent on who was selected in Event 1. If, for example, a member of Party A is selected in Event 1, the probability of selecting a member of Party A in Event 2 is 0.44 (4 / 9 = 0.44), and the probability of selecting a member of Party B in Event 2 is 0.56 (5 / 9 = 0.56).

Conditional probabilities reflect the conditional dependency of probabilities. They are used when the event outcomes are not independent. In the political party example, the probability of selecting a member of Party A in Event 2 is conditional on what happened in Event 1. If we know that a member of Party A was selected without replacement in Event 1, the probability of selecting a member of Party B in Event 2 is 5 out of 9, or 0.56. So the behavior of the subsequent events needs to be given to determine later event probabilities when the events are not independent. We can express the conditional probabilities of the two-event experiment example as follows:

P(of selecting Party B member, given that we selected Party A member in Event 1) or P(B|A)

The expression of P(B|A) is read, "The probability of B given A," or "The probability of selecting B in Event 2, given that A was selected in Event 1." Conditional probabilities are used to adjust the rule of multiplication to permit the possibility of event outcomes not being independent. When the probabilities are not independent, we should use the following rule of multiplication:

Rule of Multiplication (for not independent events A and B):

P(A and B) = P(A) × P(B|A)

To illustrate this rule, consider the voter selection problem. What is the probability that a member of Party A will be selected first, without replacement, and then a member of Party B? In this experiment, there are two events whose outcomes are not independent, because the probability of Party A being selected in Event 1 will alter the probability of Party B being selected in Event 2. The probability of a member of Party A being selected in Event 1 is P(A) = 0.5, and the probability P(B|A) is 0.56 (5 / 9 = 0.56). So the probability of selecting a member of Party A and a member of Party B follows:

$$
\begin{aligned}
P(A \text{ and } B) &= P(A) \times P(B|A) \\
&= 0.5 \times 0.56 \\
&= 0.28
\end{aligned}
$$

This probability means that there is a 28 percent chance that a member of Party A will be selected from the sample of ten, and then a member of Party B will be selected from the remaining sample of nine members.

Question: What is the probability of three members of Party A being selected, without replacement, in the sample of ten members mentioned earlier?

Answer:

$$P(A_1 \text{ and } A_2 \text{ and } A_3) = P(A_1) \times P(A_2|A_1) \times P(A_3|A_2)$$
$$= (5/10) \times (4/9) \times (3/8)$$
$$= 0.09$$

A.4 Probability Distributions

There are several basic terms that are used to describe probabilistic situations and distributions: experiments, trials, outcomes, and events. An *experiment* is defined as a test or a *trial,* or a set of tests or trials, in which an operation is conducted to discover unknown behavior. A consumer survey is an experiment that can be used to determine unknown consumer behavior. Each time a meal is prepared in a restaurant for a customer, it represents an experiment of the chef trying to satisfy the customer's tastes. An *outcome* is defined as the result of the experiment. The results of a consumer survey or the preparation of a meal can have a successful or unsuccessful outcome. An *event* is also defined as the outcome of an experiment. If an experiment consists of only one trial, the results of the experiment can be called an outcome or an event of the experiment. Events usually represent the different types of outcomes that are possible in an experiment when the experiment consists of multiple tests or trials. For example, in a product failure study, products are tested to determine the probabilities of a product working or failing. In this two-outcome situation, the probability of a product working is one event, whereas the probability of a product failure is another event.

We can define a *random variable* as any value resulting from a random experiment that by chance can generate different values. *Probability distributions* are described in the form of a graph, table, or formula, the probable behavior of a random variable's events in an experiment. There are two types of random variables, and two types of probability distributions to describe them. A *discrete random variable* is one whose events are integer values, starting with 0, 1, 2, and so on to some positive value that can be counted. As a trial is repeated and its events recorded into a frequency distribution, it can become a probability distribution. A *discrete probability distribution* specifies the probability associated with each possible event of the discrete random variable. All discrete probability distributions have two characteristics: the probabilities of the random variables are greater than or equal to zero, and the summation of all the random variable's probabilities is equal to one. A *continuous random variable* is one whose events can assume any integer or noninteger (real number) value. Any values that can be measured in decimals such as height, weight, or length are values that fall within an interval that can be considered continuous random variables.

Question: How many possible continuous variable weight values are there between 150 and 151 pounds?

Answer: An infinite number depending on how precise the values need to be.

The probability distribution for a continuous random variable is called a *continuous probability distribution*, a *probability density function*, because it describes how the probability is distributed within the function of the random variable, or a *frequency function*, because it originates from frequencies. The continuous probability distribution specifies the probability associated with each possible event or interval of event occurrences for the continuous random variable. Like a discrete probability distribution, a continuous probability distribution has two characteristics: the probabilities of the random variables are greater than or equal to zero, and the summation of all the random variables' probabilities is equal to one. In addition, the continuous probability distribution assumes that the range of events is infinite.

What follows will primarily present probability distributions as a means of obtaining probability statistics that describe an experiment or business decision-making situation consistent with the descriptive step of BA. It should also be mentioned that probability distributions provide the foundation for most of the inferential or predictive statistical methods discussed in other steps of the BA process. SAS uses basically the same approach for the assessment and calculation of probability distribution values. This approach involves defining the type of probability distribution and then placing the needed parameters in the software so that probability estimation can be calculated. The probability distributions in this section will be illustrated using SAS.

A.4.1 Discrete Probability Distribution

This section examines a series of discrete probability distributions, including the binomial, Poisson, geometric, and hypergeometric. For each probability distribution, its formula and parameters are presented (to understand the need for the model's parameters), along with the distribution characteristics (to identify where the probability distribution should be applied), and an example is given (to show how to obtain the probabilities, know what information they provide, and observe some computer-generated solutions).

A.4.1.1 Binomial Probability Distribution

The *binomial probability distribution* is a discrete probability distribution in which the outcome of an experiment is limited to two events, like success or failure, yes or no, true or false. To simplify the two-outcome nature of this distribution, the generally accepted "success" or "failure" outcomes will be used in describing the two-outcome distribution.

The binomial probability distribution formula used to calculate the probability of a success is

$$P(r) = \frac{n!}{r!(n-r)!}(p)^r(q)^{n-r}$$

where:

r = exact number of successes observed or desired (the random variable)

$P(r)$ = probability of the exact number of successes occurring

n = number of trials in the experiment

p = probability of a success, r, on each trial

q = probability of a failure on each trial, or $q = 1 - p$, because $q + p$ must equal 1 in a two-outcome trial

The characteristics of this binomial probability distribution include the following:

- In a single trial or outcome of an experiment, there can be only two outcomes.
- The two outcomes for any of the n trials are mutually exclusive.
- The probability of r remains constant for each trial in the n trials of the experiment.
- The n number of trials is independent, so the result of one trial does not affect the outcome of any other trial.
- The data used to establish the distribution is based on counting all the n possible trial results.
- The binomial probability distribution tends to be positively skewed when the probability of success, $P(r)$, is less than 0.5 and the number of n trials is fairly small (the number of trials is close to zero).

Question: Suppose we have just ordered 100,000 light bulbs. We want to test the quality of the bulbs to see if they will work before we accept them from the manufacturer. We have purchased the light bulbs with the agreement that up to 10 percent of the bulbs can be defective before rejecting the lot. To check out the quality of the lot, we take a sample of 10 light bulbs (i.e., $n=10$) to test. We will test all 10 light bulbs, and if either 0 light bulbs ($r = 0$) or 1 light bulb ($r = 1$) is defective, we will still accept the entire lot of 100,000. If 2 or more light bulbs are defective and fail to light from the sample of 10, the entire lot will be rejected. Based on past experience with hundreds of thousands of light bulbs, we know that the probability of any one bulb being defective and failing to light is only 0.05 ($p = 0.05$); the probability of any single bulb being nondefective and lighting on the first try is 0.95 ($q = 0.95$). What is the probability of 0 or 1 bulb (10 percent or less in the sample of 10 bulbs) being defective?

Answer: To successfully use the binomial probability distribution as is the case in other probability distributions, identify the distribution's variable (r) and the parameters (n, p, and q), and match the characteristics of the problem with the binomial probability distribution characteristics. Reviewing the six binomial probability distribution characteristics in this light bulb problem, one can see how this problem fits this type of distribution. Remember that finding a defective light bulb is an r *outcome*; the probability of finding a defective light bulb, p, is 0.05; each time a light bulb is tested it is a *trial*; the *experiment* consists of ten light bulb tests or trials. The characteristics follow:

- For any single light bulb test, there are only two outcomes: a defective or a nondefective light bulb.
- The defective or nondefective outcomes in any test trial are mutually exclusive.
- The probability of finding a defective light bulb r remains constant at 0.05 for each test trial in ten tests of the experiment.
- The ten test trials are independent, so the results of one test do not affect the outcome of any other test.
- The data used to establish the probability distribution is based on counting the ten test combinations taking $r = 0, 1, 2$, to 10 at a time.
- The resulting binomial probability distribution for the light bulb test is presented in Figure A.2 and is positively skewed because the probability of success is $p = .05$ and the number of trials ($n = 10$) is fairly small and close to zero.

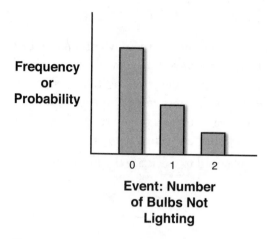

Figure A.2 Binomial probability of light bulbs not lighting

In this problem, we are trying to find out the probability of not just one event occurring, but two events ($r = 0$ and $r = 1$). Remember, when there are two mutually exclusive events occurring, add their probabilities together to determine their union of probable occurrence. This is accomplished quite easily with the binomial probability distribution software functions in SPSS and Excel. Simply compute the probability of P($r = 0$), where $p = 0.05$ and $n = 10$, and then add it to P($r = 1$), where $p = 0.05$ and $n = 10$ using the software functions. Using the SAS function, one can automatically compute the cumulative probabilities of P($r = 0$ or $r = 1$), as shown in Figure A.3. As we can see in the figure, the probability of 0 or 1 light bulb being defective is P($r = 0$ or $r = 1$) = 0.9138616441. Cumulative probabilities are commonly used in business applications.

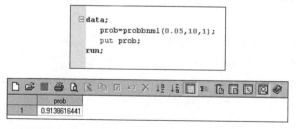

Figure A.3 SAS binomial probability of light bulbs not lighting

A.4.1.2 Poisson Probability Distribution

The Poisson probability distribution is like the binomial except when the number of trials is very large and the probability of success is very small. The Poisson distribution is particularly useful when one is interested in determining the probability of some number of events occurring during a specific time period (for example, the probability of exactly four people waiting in line for a service) or in a specific area (for example, the probability of one or more paint blemishes occurring on a product).

The Poisson probability distribution formula used to calculate the probability of a random event occurring follows:

$$P(X) = \frac{e^{-\mu}\mu^x}{X!}$$

where:

X = exact number of successes observed or desired

$P(X)$ = probability of the exact number of successes occurring

e = the constant value of 2.71826

μ = mean number of X successes occurring per unit of time, area, volume, and so on

The characteristics of this Poisson probability distribution include the following:

- The random variable X must be a nonnegative integer.
- The experiment consists of counting the number of times a single event occurs for a given time period, in an area, or in some volume. Within the time, area, or volume of the experiment, there is a large number of possible event occurrences, and the probability of any event is small.
- The number of events in one unit of time, area, or volume is independent of the number of any other units.
- The mean or expected number of success occurrences μ must remain constant for the same time period in an area or in some volume.
- The Poisson probability distribution tends to be positively skewed because μ must remain constant for the same time period in an area or in some volume.

As with the binomial distribution, to successfully use the Poisson probability distribution, identify the distribution's variable (X) and the parameter (μ) and match the characteristics of the problem with the Poisson probability distribution.

Question: Suppose we are planning on purchasing a printer to handle computer jobs that need to be printed. The size of the jobs and the flow to the printer is Poisson distributed with a mean arrival rate of two jobs per minute ($\mu = 2$). Model A has temporary storage areas to hold computer jobs until the printer can handle them. The temporary storage areas are called *buffers*. Unfortunately, the way the computer system is set up, if the job does not go into buffered storage, it has to be stored in the main computer at an almost prohibitive cost. The selection of the printer will be based in part on its probable ability to handle the flow of jobs from the computer without requiring costly computer storage. Model A can buffer up to five jobs at a time. What is the probability that Model A will have to store exactly three jobs in its buffer storage per minute?

Answer: Again, it was necessary to make a number of assumptions to fit the Poisson characteristics of this problem. To the extent that the assumptions are accurate, so are the statistics based on them. We can review the first five of the Poisson probability distribution characteristics in this computer problem so we can see how this problem fits this type of distribution. Remember that the *random variable X* is the number of jobs to store in the printer in a minute; the *mean arrival rate* μ is two jobs per minute; and the *experiment* consists of counting the number of jobs that occur in a minute. The characteristics are as follows:

- The random variable X is an integer value of 0 jobs, 1 job, 2 jobs, and so on, to store in the printer's buffer storage.
- The experiment consists of counting the number of jobs arriving in a minute, and because there could be an infinite number of jobs arriving, the probability of any job arriving could be assumed to be infinitely small.
- The probability of any job arriving can be assumed to be equal to any other job arriving.
- The arrival of jobs in one minute is assumed to be independent of the number of jobs arriving any other minute.
- The mean number of jobs arriving μ is assumed constant for any minute.

Using the SAS function 'POISSON', we can automatically compute the cumulative probabilities of P(X = 3), as shown in Figure A.4. As we can see in the figure, the probability of P(X = 3) = 0.1804470443.

```
data;
    prob=pdf('POISSON',3,2);
    put prob;
run;
```

	prob
1	0.1804470443

Figure A.4 SAS Poisson probability of three print jobs in the buffer

A.4.1.3 *Other Discrete Probability Distributions*

There are many other discrete probability functions. Two additional probability distributions commonly used in business analytics include those in Table A.1. These can also be computed using SAS function statements.

Table A.1 Other Discrete Probability Distributions

Probability Distribution	Description	Application Examples
Geometric	Similar to the binomial in that there is a two-outcome situation of success r or failure, the p probability of success occurring, and q probability of failure of the event not occurring. Unlike the binomial, one is interested in the first success that will occur out of some unknown number of n trial periods. This distribution is particularly useful when the interest is in determining the probability of some events occurring in a specific number of discrete time periods (for example, in hours, days, minutes).	Examples of the use of this distribution can include determining the probability that a customer will wait two or more minutes in a queue for service; the probability that out of four people surveyed, one will buy a new product; and the probability that one will have to sample at least 20 tax returns before finding the first person who will need to be audited.

Probability Distribution	Description	Application Examples
Hypergeometric	Similar to the binomial distribution in that it is a two-outcome situation of success *r* or failure, but with this distribution the trials are dependent. The dependence of the trials occurs because the results are taking place *without replacement*. In using this probability distribution, unlike all other distributions discussed so far, a total number or *population* of elements must be specified, denoted *N*, from which sampling takes place without replacement. This distribution is based on counting and does not require known probability parameters in its computation.	Examples of the use of this distribution include determining the probability of lot rejection based on lot size and sample size, the probability that a person has been discriminated against based on population size, and determining the proportion of voters that will vote for a specific political party's candidate based on a sample of registered voters.

A.4.2 Continuous Probability Distributions

Two of the most common continuous probability distributions are the normal and the exponential. With these distributions, we compute the area under the curve between two points or from one point out to infinity (the tails of the distribution).

A.4.2.1 Normal Probability Distribution

The most commonly used probability distribution is the *normal probability distribution*. It is a continuous probability distribution, where the outcomes of an experiment are expressed as a continuous function. This distribution can be applied in almost any problem situation in which the event in an experiment follows a continuous function and meets the distribution's characteristic requirements.

The normal probability distribution formula used to calculate the probability of the function of *X* for a specific measurable random variable *X* value follows:

$$f(X) = \frac{1}{2\pi} e^{-\frac{(X-\mu)^2}{2\sigma^2}}$$

where:

X = a measurable normal random variable

f(X) = probability of exactly X occurring (usually referred to as a function of X)

σ^2 = variance of the normal random variable

µ = mean of the normal random variable

π = a constant of 3.1416

e = a constant of 2.71828

Unlike the discrete distributions, the characteristics of the normal probability distribution are more directed at describing the distribution, rather than an experiment that the distribution will be used in. The characteristics of the normal probability distribution include these:

- The mean occurrence of the measurable random variable *X* is µ, with a standard deviation of σ.
- The random variable *X* ranges over ±∞. (This characteristic is a theoretical requirement that is, of course, not observed in practical applications.)
- The shape of the curve is *bell shaped*, like that in Figure 5.12 in Chapter 5, "What Is Descriptive Analytics?"
- The distribution is evenly divided by its mean value of µ.
- The distribution is *symmetrical*, with an equal number of values falling on both sides of µ.
- The distribution is *asymptotic*, so the tails of the curve drop off toward the horizontal axis but never touch it.

With a continuous function, we are interested in probabilities and usually not the value of the normal function f(*X*). The value of f(*X*) on a continuous function would represent a very small probability. That is, the probability of any specific value of *X* occurring out of the infinite values of *X* has an infinitely small probability of occurrence. To determine the probabilities of the normal probability distribution, we convert the area under the continuous function into a probability space, as presented in Figure A.5.

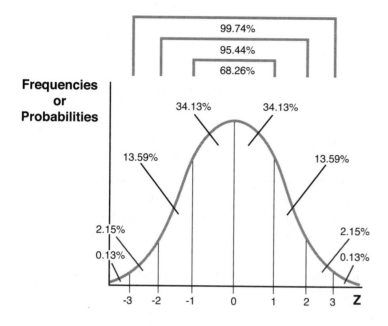

Figure A.5 Normal distribution with Z values

The actual determination of probabilities can be difficult even though a normal distribution fits the preceding characteristics. Unfortunately, the shape of the curve and the distribution's parameters impact probability calculations. Every distribution found in the real world might qualify for a normal distribution but would have differing areas under the curve (more or less skewedness). To deal with the variety (or sometimes called *family*) of normal probability distributions, mathematicians developed a single distribution called the *standard normal probability distribution* (see Chapter 5, Section 5.5). The characteristics for this distribution are the same as the normal probability distribution. The standard normal probability distribution formula used to determine the probabilities for any normal probability distribution follows:

$$Z = \frac{X - \mu}{\sigma}$$

where:

Z = the number of standard deviation units between X and μ

X = a specific random variable value

μ = mean of the normal random variable

σ = standard deviation about μ

The standard normal probability distribution has a mean of 0 and a standard deviation of 1. That is, μ is set equal to zero, and we can add and subtract units of σ to denote areas or probability spaces in the distribution. The measurable variable X is converted into units of standard deviation called Z *value*. As seen in Figure A.5, the area under the normal curve covering $\pm 1\sigma$ is the same as a Z value of ± 1. The area under the normal curve of a Z value ± 1 represents 68.26 percent; in other words, the probability that X will occur between $\pm 1\sigma$ is 0.6826.

Question: What is the probability that X will fall between $\pm 2\sigma$ in a normal probability distribution?

Answer: The probability can be seen in Figure A.5 as 0.9544, or 95.44 percent.

The Z values provide a standardized measure of deviation units. Because all normal probability distributions have a standard deviation, the Z values provide a common or standardized measure by which all normal probability distributions can be linked. Once the standardized normal probability distribution values are determined, they can be used as a standard to approximate probabilities for all the other normal probability distributions. Indeed, the purpose of the standard normal probability distribution is to provide the probabilities for all normal distributions, given a Z value.

Question: Suppose we use an automated piece of equipment to fill metal cans with a beverage. The metal can that the machine fills is designed to hold as much as 14.5 ounces of beverage, but the plan is to put only 12 ounces in a metal can. After observing the first 1,000 cans filled by the machine (representing the population of the machines' can fill activities), we find that the mean number of ounces is 12 (μ), with a standard deviation of 0.5 ounces (σ). Based on this information, what is the probability that the amount of beverage filled in a can will be between 12 and 13 (the random variable) ounces?

Answer: We will assume that this problem fits the normal distribution. The probability distribution of this problem is presented in Figure A.6. For this question, determine the shaded area under the curve presented in the graph. To do this, use the Z formula to compute the Z value representing the area between $\mu = 12$ and a specific value of X, or X = 13. The Z value is 2 (13 – 12 / 0.5). So what is needed is to determine the probabilistic conversion of a Z value of 2 standard deviations into the area under the curve or a probabilistic occurrence of the X falling between 12 and 13 ounces ($P(12 \leq X \leq 13)$).

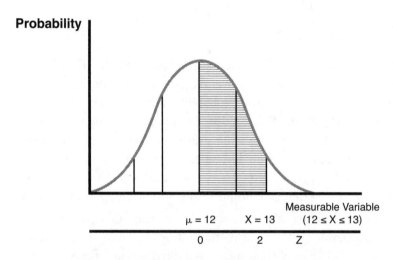

Figure A.6 Normal distribution example with Z value

Using SAS's 'NORMAL' function, we can compute this probability. This function provides cumulative probabilities from some X point to the origin (all the area to the left of the distribution from random variable X). Using this function, the Z value is the only parameter needed, but it requires two steps. First, to find the cumulative probability from the Z value of 2 where X = 13, and second, to find the probability from the Z value of 0 where X = 12. The resulting probability from the Z value of 2 as presented in Figure A.7 is 0.9772498681. This value includes all the probabilities to the left of μ, which is 0.5, or half the area under the curve. That area is where X = 12, so the resulting probability that a can will be filled between 12 and 13 ounces is 0.4772498681 (0.9772498681 – 0.5).

```
data;
    prob=cdf('NORMAL',2);
    put prob;
run;
```

	prob
1	0.9772498681

Figure A.7 SAS's computation of normal probability example

A.4.2.2 *Exponential Probability Distribution*

The exponential probability distribution is a continuous probability distribution, where the values of the variable X are measurable and expressed as a continuous function. This distribution is similar to the Poisson probability distribution in that it deals with time in the distribution. Unlike the Poisson distribution, though, time is not fixed but is the actual variable. The exponential probability distribution has been extensively used in queuing analysis to estimate the probable time required of a service facility to process a customer. In a similar way, the exponential probability distribution is used to determine computer processing time of computer software by computer hardware.

The exponential distribution formula is expressed as a continuous function and will be present in a queuing context of measuring the timing of units arriving at a service facility. The continuous function of the exponential distribution follows:

$$f(X) = \lambda e^{-\lambda a}$$

where:

X = a measurable random variable representing the time between successive arrivals to the service facility

$f(X)$ = value of the function at X

a = the specific value of X whose probability is being sought

λ = a Greek letter, *lambda,* representing the average rate of arrivals to a service facility (like the µ of the Poisson distribution)

e = a constant of 2.71828

We calculate the area under the curve and X between a and $\pm\infty$ to determine the probability of the function. Here's the formula for determining the probability of a or more units being served:

$$P(\geq a) = e^{-\lambda a}$$

where:

$P(X \geq a)$ = probability of $X \geq a$ occurring

Following are the characteristics of the exponential probability distribution:

- The mean arrival rate of the measurable random variable X is λ.
- Both X and λ must be positive, and a must be within the range of X.
- The random variable X ranges from 0 to $+\infty$. (This characteristic is a theoretical requirement that is, of course, not observed in practical applications.)
- The experiment consists of determining the probability of X having a or more arrivals when the mean arrival rate of λ is known and continuous.
- The shape of the curve tends to be positively skewed (see Figure A.8).

For this continuous exponential probability function, the focus is on the probabilities and usually not the value of the exponential function $f(X)$. To determine the probabilities of the exponential probability distribution, convert the area under the continuous function into probabilities as with the normal distribution. To successfully use the exponential probability distribution, identify the distribution's variable (X) and the parameter (λ), and match the characteristics of the problem with the exponential probability distribution characteristics.

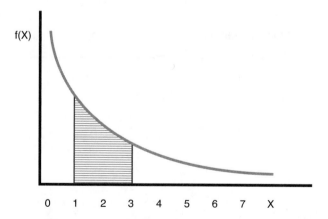

Figure A.8 Exponential probability distribution example

Question: Suppose we have been trying to find out the probable processing time for jobs on a computer. Most jobs require little or almost no processing time, but a few jobs do take more time. From internal processing records, we know that the mean processing rate is one job per microsecond (λ = 1 job per microsecond). What is the probability that a computer job will take from one to three microseconds, inclusively?

Answer: Assume that this problem meets the characteristics of the exponential probability distribution. The shaded region in Figure A.8 is the probability area that needs to be determined. Using SAS's 'EXPO' function, we only need to determine the distribution's variable (X) and the parameter (λ). The random variable is an interval value $P(1 \leq X \leq 3)$ and $\lambda = 1$. The 'EXPO' function provides cumulative probabilities (from 0 out to X in Figure A.8), as does the 'NORMAL' function. This means the calculation requires first determining $P(X = 1)$ and subtracting it from $P(X = 3)$. The SAS exponential probability calculation for $P(X = 1)$ is presented in Figure A.9 as 0.63212 (rounded in this illustration). It turns out that $P(X = 3) = 0.95021$, resulting in $P(1 \leq X \leq 3) = 0.31809$. So the probability of a computer program taking from 1 to 3 microseconds is 31.809 percent.

```
data;
    prob=cdf('EXPO',1);
    put prob;
run;
```

	prob
1	0.6321205588

Figure A.9 SAS exponential probability distribution example

A.5 Statistical Testing

In Chapter 5, when confidence intervals were discussed, the focus was on creating boundaries when a population parameter was expected to fall, given a specific confidence level (the number of plus or minus Z values and the related percentages; see Table 5.6 in Chapter 5). As can be seen from the two-tailed normal distribution in Figure A.10, a 95 percent confidence level is computed from a Z value of 1.96.

We are assuming use of a confidence coefficient such that the true population parameter will fall within the confidence interval. What if it actually falls outside the interval? That is the question that hypothesis testing helps determine. In a scientific and formal manner, all hypothesis tests have a *null hypothesis* (designated as H_0). H_0 implies that there is no significant difference between the mean that we have from our sample (or some other measure of central tendency) and the population mean. The interval in Figure A.10 also means that the 95 percent confidence allows a 5 percent chance the population mean is not in the interval. That 5 percent is called the *level of significance*, because it marks the boundary of where one designs a significant difference between the sample mean and a population mean. The level of significance is often labeled with a small alpha, α. So in Figure A.10, a 95 percent confidence level will have a 5 percent level of significance. If it turns out that the sample mean is significantly different from the expected population mean, we mark the occasion by rejecting the H_0 and accepting the *alternative hypothesis*, H_1. H_1 can be an inequality, a greater-than or equal-to or a less-than or equal-to expression depending on what is being tested. Note in Figure A.10 that there can be a two-tail test that is for the inequality expression and a one-tail test that is for greater-than or equal-to expressions.

A Two-Tail Test

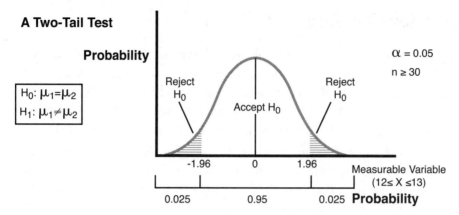

A "Less-Than" Alternative Hypothesis One-Tail Test

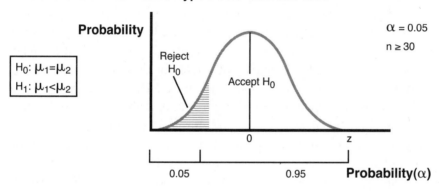

A "Greater-Than" Alternative Hypothesis One-Tail Test

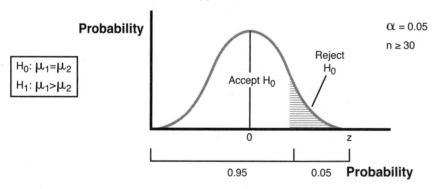

Figure A.10 Hypothesis testing, one- and two-tailed tests

How does one determine if there is a significant difference between the random variable, which can include means, proportions, variances, or any kind of measure of central tendency computed from a sample, and the confidence interval value boundaries? These tests are accomplished through a variety of statistical tests (see Table A.2). When the statistic being compared is a population parameter based on a probability distribution, the hypothesis test is called a *parametric hypothesis test*. Alternatively, for hypothesis tests with measures that are not population parameters (based on counts or frequencies), use a *nonparametric hypothesis test*. There are many of both hypotheses tests in the literature. SAS offers an impressive array of these tests, many of which are built into functions that can provide a level of significance on which statistics can reveal useful information for a BA analysis.

Table A.2 Common Parametric Statistical Tests and Software Access Information

Test Statistic	Application Area	Access to SPSS Function	Access to Excel Function
F-Test Two Sample for Variances	Compares the variances from two samples to see if they are from the same probability distribution	Analyze > Compare Means > One-Way ANOVA	Data Analysis > F-Test Two Sample for Variances
t-test: Paired Two Sample Means	Compares the mean values from two samples to see if they come from the same probability distribution	Analyze > Compare Means > Paired Samples t-test	Data Analysis > t-test: Paired Two Sample Means
Z-test: Two Sample Means	Compares the mean values from two populations to see if they have the same probability distribution	For sample sizes above 30: Analyze > Compare Means > Paired Samples t-test	Data Analysis > Z-test: Two Sample Means
ANOVA: Single Factor	Compares the variance between and within two or more samples to see if the samples are drawn from the same probability distribution	Analyze > Compare Means > One-Way ANOVA	Data Analysis > ANOVA: Single Factor

To illustrate the use of one of the parametric statistical tests, use the sales data presented in Chapter 5 (Figure 5.1), which is repeated here in Figure A.11. To compare the mean sales data for Sales 1 with that of Sales 2 to determine if they are closely enough distributed, use either one for BA predictive purposes. In other words, it is important to determine if they are from the same probability distribution. Assume these values are paired with each other for this example. Because the data are samples, use the t-test rather than the Z-test because the Z-test is used to make comparisons using population-sized data.

```
DATA sales_data;
  INPUT obs sales1 sales2 sales3 sales4 ;
CARDS;
1 23 1234 1 1
2 31 943 2 5
3 48 896 3 9
4 16 12 4 12
5 28 15 5 18
6 29 15 6 19
7 31 23 6 19
8 35 21 6 21
9 51 25 6 21
10 42 27 7 21
11 34 27 8 21
12 56 29 9 21
13 24 20 10 21
14 34 18 11 19
15 43 13 12 19
16 56 8 13 18
17 34 7 14 12
18 38 6 15 9
19 23 4 16 5
20 27 1 17 1
;
RUN;
```

Figure A.11 Illustrative sales data sets

Using the appropriate access to the software functions (Table A.2) and entering the data from Figure A.11 for Sales 1 and 2, the software can compute the necessary test statistics. Because we are not looking specifically for one mean to be larger than the other and direction does not matter, we can use a two-tail test. The resulting SAS (Table A.3) test statistics are presented next. SAS software computed the t-test statistic as −1.57. (Note with a two-tail test, either negative or positive values can occur.) This is then compared to the related area under the curve in the probability distribution in terms of units of deviation. The result is that the software computes the resulting significance level, which is $p = 0.1322$. The p symbol is often used when the exact test probabilities are computed, as opposed to the desired significance level of alpha, α. If a set desired significance level is 5 percent ($\alpha = 0.05$), it could be concluded that there is no significant difference in the paired comparison of the mean values from the same distribution of sales. So formally, we would accept H_0 and, therefore, accept

H_1 if this example was formally structured as a hypothesis test. To have concluded there is a significant difference between the means, the resulting level of significance, p, would have had to have been less than 0.05. So for any of the test statistics, the computed level of significance is an essential test statistic result used to judge the outcome of the statistical test. The greater the computed level of significance, p, the less significantly different the parameters are assumed to be.

```
proc ttest data=sales_data;
   paired sales1*sales2;
   run;
```

```
                    The TTEST Procedure

               Difference:   sales1 - sales2

 N        Mean      Std Dev      Std Err     Minimum      Maximum
20       -132.1      375.4       83.9362     -1211.0      48.0000

    Mean          95% CL Mean          Std Dev        95% CL Std Dev
  -132.1        -307.7   43.6306        375.4          285.5    548.3

                    DF      t Value     Pr > |t|
                    19       -1.57       0.1322
```

Table A.3 SAS T-Test Statistics for Sales Example

Unlike the Z-test, t- and F-tests do not use means (a parameter) in their computed value computations. These types of hypothesis tests can be grouped into what is descriptively called *nonparametric tests*, or *distribution free tests*. Table A.4 lists some of the more common of these statistical tests.

Table A.4 Common Nonparametric Statistical Tests and Software Access Information

Test Statistic	Application Area
Binomial Test	This test compares the observed frequencies of the two categories of a dichotomous variable to the frequencies that are expected under a binomial distribution with a specified probability parameter (α).
Chi-Square	This test (used often for a goodness-of-fit test) compares the observed and expected frequencies in each category of a distribution to test that all categories contain the same proportion of values or to test that each category contains a user-specified proportion of values.
Kolmogorov-Smirnov (One-Way)	Multiple versions of this test can be applied to differing comparative analyses. The one-way procedure compares the observed cumulative distribution function for a variable with any specified theoretical distribution.

Test Statistic	Application Area
Wilcoxon Signed-Rank	Multiple versions of this test can be applied to differing comparative analyses. This test is applicable when two samples (two populations) are related (not independent). The test is designed to compare some n matched pairs of ranked or ordinal data from two populations.
Run	A *run* is a sequence of like observations. This tests whether the order of occurrence of two values of a variable is random. A sample with too many or too few runs suggests that the sample is not random.

Each software system provides a differing number of nonparametric tests. SAS has functions that provide the Chi-Square test statistic value and access to all the nonparametric tests in Table A.4. This particular set of functions helps the BA analyst by allowing the SAS software to determine the best test to select for the analysis.

B

Linear Programming

B.1 Introduction

Linear Programming (LP) is a deterministic, multivariable, constrained, single-objective, optimization methodology. It's a model with known, deterministic, and constant parameters, and it has more than one unknown or decision variable. LP has mathematical expressions that constrain the values of the decision variables, and it seeks to solve for an optimal solution with a single objective. It is a general-purpose modeling methodology, permitting application to just about every possible problem situation that fits the assumptions the model requires. (We will discuss the assumptions of the LP model in a later section of this appendix.) Specifically, LP can be used to model problems in all the functional areas of business (accounting, economics, finance, management, and marketing) and in all types of operations (industry-wide, government, agriculture, health care, and so on).

Modeling a problem using LP is called programming. As such, LP is considered one of several mathematical programming methodologies available for use in the prescriptive step of the business analytic process.

B.2 Types of Linear Programming Problems/Models

There are basically two types of LP problems/models: a maximization model and a minimization model. Business seeks to maximize profit or sales. In such cases, the single objective is maximization. Other business situations seek to minimize costs or resource utilization. In those cases, the single objective is minimization.

In addition to these two basic types of LP models, there is a group of special case models. These models are also maximization or minimization models, but they are applied to a limited set of problems. One example is integer programming (discussed in Appendix D, "Integer Programming"), whose model solution requires integer values rather than real number solutions.

B.3 Linear Programming Problem/Model Elements

B.3.1 Introduction

All LP problem/model formulations consist of three elements: an objective function, constraints, and nonnegativity or given requirements. The generalized model (a model without actual values, only symbols) requires the three components presented in Exhibit A. Note that the applied model is also presented in Exhibit B. Both models will be discussed in this section. The exhibit used here foreshadows the formulation of models discussed in this appendix.

A. Generalized LP Model

Maximize: $Z = c_1 X_1 + c_2 X_2 + \ldots + c_n X_n$ (Objective Function)

subject to: $a_{11} X_1 + a_{12} X_2 + \ldots + a_{1n} X_n \le b_1$ (Constraints)

$a_{21} X_1 + a_{22} X_2 + \ldots + a_{2n} X_n \le b_2$

.

.

.

$a_{m1} X_1 + a_{m2} X_2 + \ldots + a_{mn} X_n \le b_m$

and $X_1, X_2, \ldots, X_n \ge 0$ (Nonnegativity or Given Requirements)

B. An Applied LP Model *(Ford Motor Company problem/To be explained in this appendix)*

Maximize: $Z = 2000 X_1 + 3500 X_2$

subject to: $60 X_1 + 75 X_2 \le 10000$

$60 X_1 + 75 X_2 \ge 3000$

$X_1 + X_2 = 140$

and $X_1, X_2 \ge 0$

B.3.2 *The Objective Function*

The objective function is generally expressed as one of the following:

Maximize: $Z = c_1 X_1 + c_2 X_2 + ... + c_n X_n$

or

Minimize: $Z = c_1 X_1 + c_2 X_2 + ... + c_n X_n$

where:

Z = an unknown that is not a variable but one that will be solved when the values of the decision variables are determined

X_j = decision variables for j = 1, 2, ..., n; which are the unknowns to solve for an optimal value

c_j = contribution coefficients for j = 1, 2, ..., n; which represent the per-unit contribution to Z for each unit of the decision variable to which they are related

The objective function is always an equality expression with the same form and style as the preceding two. In this book, the coefficients are always positive (although in some real-world problems, they can be negative). If in a problem the single objective is maximizing profit, use the Maximize Z function. If the objective is to minimize costs, use the Minimize Z function.

This objective function can be illustrated by a simple problem. Suppose one wants to decide how many automobiles a Ford Motor Company plant should produce in a week. The plant is capable of producing only two types of automobiles: Mustangs and Thunderbirds. So the decision variables in this LP model will be as follows:

X_1 = number of Mustangs to produce per week
X_2 = number of Thunderbirds to produce per week

The plant would not produce automobiles unless it could make some profit from the endeavor. Suppose it could make \$1,000 on each Mustang and \$3,500 on each Thunderbird. These values (1,000 and 3,500) represent the per unit of automobile profit contribution to what will be the total profit (Z) and are the contributions coefficients c_1 and c_2 in the model. The resulting objective function for this problem would be this:

Maximize: $Z = c_1 X_1 + c_2 X_2$ (generalized form)
Maximize: $Z = 1000 X_1 + 3500 X_2$ (applied form)

If this objective function had no constraints to limit the size of the decision variables, they could be set at positive infinity to make as much profit as possible. Unfortunately, in the real world, there are always constraints to limit the optimization effort.

B.3.3 Constraints

The constraints in an LP model can generally be expressed as the following:

subject to:

$$a_{11} X_1 + a_{12} X_2 + ... + a_{1n} X_n \leq b_1$$
$$a_{21} X_1 + a_{22} X_2 + ... + a_{2n} X_n \geq b_2$$
$$...$$
$$a_{m1} X_1 + a_{m2} X_2 + ... + a_{mn} X_n = b_m$$

where:

b_i = a right-hand-side value for i = 1, 2, ... m; and "m" is the number of constraints in the model, each having a right-hand-side value usually representing a total resource availability or requirement

a_{ij} = technology coefficients for i = 1, 2, ..., m and j = 1, 2, ... n; which represent the per-unit usage of the related ith right-hand-side value by the related jth decision variable

In the constraints, the technology coefficients (a_{ij}) are located by row with the first subscript and by column with the second subscript. The term *technology coefficient* is used to describe this parameter because technology applications are the principle determiner to the size of this coefficient.

LP constraints come in only three expressions: ≤, ≥, or =. Some models have only one type of expression for all their constraints; other models can use all three types. How does one know when a particular type should be used? It depends on the related right-hand-side b value. If b is a total maximum value (like total number of labor hours that at most can be used for production), then use a ≤ expression. If b is a total minimum value (like total minimum number of labor hours that are contracted for production), then use a ≥ expression. If b is an exact value (like a jeweler who has 20 diamonds and must use exactly 20 diamonds in 20 necklaces), use an = expression.

The left-hand-side of the constraint represents resources that produce the decision variable values. The right-hand-side (RHS) represents the number of resources to be considered in the model. When the model solves for the optimal decision variable values, they have to conform to the limitations posed by these constraints. This is why the constraints begin with the two words "subject to." The objective function is subject to (or limited by) the constraints.

How many constraints are enough for modeling a problem? The answer depends on the problem. In this appendix, the formulation of constraints is based on available information from word problems. In an actual real-world problem, modelers are guided by available information or data. Like a painter placing paint on a canvas, a modeler adds as many constraints in the model as there is available data to formulate them. LP is a robust model that eliminates or makes redundant constraints that are not needed in most cases. But there is a balancing effort that must be considered. On the other hand, as can be seen in a later section on model formulation complications, too many constraints can spoil the formulation if they are formulated incorrectly. Too few constraints also cause another complication preventing an accurate model from being formulated. Too many or too few are both examples of incorrect modeling, and it is only through experience that modelers can learn to successfully formulate LP models. The experience provided in this appendix to practice the formulation of these constraints will increase model formulation skills in determining how many or how few constraints should be included.

Continuing with the Ford Motor Company problem, in an effort to maximize profit, there are some weekly resource limitations for production. First, there are only 10,000 hours of skilled labor at maximum use in the production of the Mustangs (X_1) and the Thunderbirds (X_2). Suppose that it takes 60 hours for each Mustang and 75 hours for each Thunderbird. To model this constraint, we again return to the generalized form:

$a_{11} X_1 + a_{12} X_2 \le b_1$ (generalized form)

So, the resulting applied first constraint for the model, given the parameters above is the following:

$60 X_1 + 75 X_2 \le 10000$ (applied form)

The value of 60 hours is the per-unit usage of the total 10,000 hours available for each unit of the related Mustang decision variable. When the optimal values for X_1 and X_2 are determined, the sum of the product with its respective technology coefficients must be less than or equal to the total maximum amount of skilled labor of 10,000 hours. Note in the constraint that there are no commas to denote the thousands. This is because model parameters will eventually be entered into software that does not accept the commas in the model formulation.

Suppose this company also has a minimum usage requirement with the skilled labor such that it must use at least 3,000 hours each week due to labor contract requirements. The second constraint for this model would then be as follows:

$60 X_1 + 75 X_2 \ge 3000$

Now suppose this company is under contract to produce exactly 140 automobiles per week to make its quota. This constraint would look like this:

$X_1 + X_2 = 140$

Note in this constraint there is an implied technology coefficient of 1 in front of each decision variable.

B.3.4 The Nonnegativity and Given Requirements

The decision variables in LP models are required to be zero or some positive value. As a formal part of the correct way of formulating an LP model (as is the case in most of the mathematical programming methods), one must add an additional statement in LP model formulations that looks like this:

and $X_1, X_2, \ldots, X_n \geq 0$

These do not represent formal constraints on the model, but a limitation on the decision variables. As presented earlier, this tells users that this model requires its decision variables to be zero or any positive value, including real numbers and fractional values.

What if one wants to produce whole units of the decision variable values (like whole units of Mustangs)? That requires the solution to generate only integer values. While the subject of "integer programming" will be presented in Appendix D of this book, this additional "given requirement" would have to be included in the model so users would know of its existence. This is done by revising the nonnegativity requirements to also include these given requirements:

and $X_1, X_2, \ldots, X_n \geq 0$ and all integer

This formal requirement is not necessary to run the model in a computer, but it is required in any formulation of a model. It is that portion of the formulation that tells users who look at the model they have to use the right kind of software (either LP or integer programming) to run and obtain a particular model. In the Ford Motor Company problem, the nonnegativity requirements that permit fractional values for the automobile production could be these:

and $X_1, X_2 \geq 0$

Now revisit the generalized model in Section B.3.3 and see how the applied Ford Motor Company problem/model formulation complies with the generalized model formulation. The generalized model coefficients and terms will be used repeatedly.

B.4 Linear Programming Problem/Model Formulation Procedure

Formulation of linear programming problems requires skill. In this section, we present a stepwise procedure useful in formulating LP problems. In addition, several practice problems and formulations are presented to help build formulation skills.

B.4.1 Stepwise Procedure

The hardest part of figuring out any word problem or any real-world problem is always the first step. This stepwise procedure is a strategy for handling any kind of LP model. Big or small, it handles them all by breaking a complex process into small, achievable steps:

1. **Determine the type of problem**—A problem has to be either maximization or minimization. If the problem only mentions making profit or sales, it is most likely a maximization problem. If the problem only mentions cost, it most likely is a minimization problem. What if a problem includes sales and cost information? Then subtract the cost from the sales and derive profit. Maximizing profit both maximizes sales and minimizes cost. The values that can be used to determine the type of problem are called the *contribution coefficients*.

2. **Define the decision variables**—Step 1 determined the type of problem by finding profit or cost contribution coefficients. The number of profit or cost contribution coefficients determines the number of decision variables because these contribution coefficients are attached to the respective decision variables in the objective function.

   There are two things to remember in defining decision variables: (1) Make clear what the decision variables are determining; (2) State any "time horizon" the problem is requiring. In the Ford Motor Company example first mentioned in Section B.3.2, the definition of the first decision variable was as follows:

   X_1 = number of Mustangs to produce per week

   This definition makes clear that the "number of Mustangs" will be produced. The definition also includes the time horizon of one week. An example of what is not acceptable in the definition of a decision variable is this:

   X_1 = Mustangs

3. **Formulate the objective function**—Because the contribution coefficients, the type of problem in Step 1, and the decision variables in Step 2 have been

identified, all that is left is to combine these into the form of an objective function, as presented in Section B.3.2.

4. **Formulate the constraints**—Introduced in Section B.3.3, this step is one of the hardest. Here are two strategies that can help: (1) Right-hand-side strategy: Look at the problem for a sentence or a column in a table that lists the available resources that the model needs to achieve. These are the right-hand-side "b" parameters. Create a column vector (a column of numbers) that will represent the "b" values in the model. Then go back and read the problem again to find the technology coefficients to finish the left-hand-side of the constraint. (2) Left-hand-side strategy: In problems with tabled values, look to see if they are technology coefficients. Take the technology coefficients and align them by row or column to form the left-hand-side of the constraints. Then go back and read the problem again to find the right-hand-side values.

5. **State the nonnegativity and given requirements**—Simply use the statement of nonnegativity given in Section B.3.4.

Now practice this formulation procedure on a series of problems. These problems range from very simple to more complex. They are designed for beginners but will prep anyone in developing LP models.

B.4.2 LP Problem/Model Formulation Practice: Butcher Problem

Problem Statement: Consider the problem of a butcher mixing the day's supply of meatloaf. The butcher has two grades of meatloaf: Grade 1 and Grade 2. The butcher needs to know how many trays of each kind of meatloaf should be made. The butcher may make whole trays or any fractional number of trays. The butcher's profit is increased by $36 for each tray of Grade 1 that is mixed, and by $34 for each tray of Grade 2. If there were no constraints, the butcher would want to make both kinds of meatloaf to maximize profit. Unfortunately, the butcher has constraints that must be considered.

- **Constraint 1**—The butcher cannot sell more than six trays of meatloaf per day.
- **Constraint 2**—Only nine hours of mixing time are available for the butcher and staff. It takes two hours to mix a tray of Grade 1 and one hour to mix a tray of Grade 2.
- **Constraint 3**—The butcher has only 16 feet of shelf space for meatloaf. Each tray of Grade 1 requires 2 feet of shelf space. Each tray of Grade 2 requires 3 feet of shelf space.

- **Formulation**—This problem clearly labels the constraints to make things easy. Remember to use the five-step formulation procedure to reduce a problem to easier and smaller steps to create the model.

1. **Determine the type of problem**—This problem only mentions profit, so it has to be a maximization problem. The two sales maximizing contribution coefficients ($36 and $34) in this model determine the type of problem.

2. **Define the decision variables**—The problem says, "The butcher needs to know how many trays of each kind of meatloaf should be made." That is one hint. An easier one is in Step 1. The two contribution coefficients mean two decision variables. Because $36 is the amount of profit on a tray of Grade 1 meatloaf, the related decision variable has to be the "number of trays of Grade 1 meatloaf to make or mix per day." Note the time horizon in the sentence, "Consider the problem of a butcher mixing the day's supply of meatloaf." The resulting two decision variables can be defined as follows:

X_1 = number of trays of Grade 1 meatloaf to make (or mix) per day

X_2 = number of trays of Grade 2 meatloaf to make (or mix) per day

3. **Formulate the objective function**—The formulation of the objective function follows easily from Steps 1 and 2. It is as follows:

Maximize: $Z = 36\,X_1 + 34\,X_2$

4. **Formulate the constraints**—Take one constraint at a time. In reading the sentence for Constraint 1 ("The butcher cannot sell more than six trays of meatloaf per day"), six is a parameter in the constraint. It has to be either a technology coefficient or a right-hand-side value. If it is a technology coefficient, it has to be directly related to an individual decision variable. If it is a right-hand-side value, it must be a total available resource. "Six" is a selling limitation on total trays, not individual trays. So it is a right-hand-side or b value. Because it is also a total maximum selling limitation, the direction of the inequality will be less than or equal to. What about the left-hand-side of this constraint? Well, what is the sum of all the trays of meatloaf? That can be expressed as the sum of both decision variables, resulting in the first constraint of the model that follows:

$X_1 + X_2 \leq 6$ (selling)

It is recommended that beginning LP modelers label their constraints with a word or two so that the modeler can remember that the particular limitation has been included as a constraint. It will also be helpful to others wanting to understand the model if the constraints are labeled with understandable terms.

In Constraint 2, the sentences are, "Only nine hours of mixing time are available for the butcher and staff. It takes two hours to mix a tray of Grade 1 and one hour to mix a tray of Grade 2." In the first sentence, "nine" is a total available mixing time limitation. So it is a right-hand-side value that represents a total maximum amount of this mixing resource resulting in a less than or equal to expression. In the second sentence "two" is attached to the Grade 1 decision variable, and "one" is attached to the Grade 2 decision variable. So these two parameters are technology coefficients. The resulting constraint follows:

$2X_1 + X_2 \leq 9$ (mixing time)

In Constraint 3, the sentences are, "The butcher has only 16 feet of shelf space for meatloaf. Each tray of Grade 1 requires 2 feet of shelf space, and each tray of Grade 2 requires 3 feet of shelf space." In the first sentence, "16" is the total available shelf space limitation. So it is a right-hand-side value that represents a total maximum amount of this shelf space resource resulting in a less than or equal to expression. In the second sentence, the "2" is attached to the Grade 1 decision variable, and the "3" is attached to the Grade 2 decision variable. So these two parameters are technology coefficients. The resulting constraint is the following:

$2X_1 + 3X_2 \leq 16$ (shelf space)

5. **State the nonnegativity and given requirements**—Because the model has only two decision variables and the problem specifically allows fractional values, all that is needed is to state the same nonnegative requirements as the basic generalized model presented in Section B.3.4 as here:

and $X_1, X_2 \geq 0$

The entire formulation of the butcher problem is again presented here:

Maximize: $Z = 36X_1 + 34X_2$

subject to: $\quad X_1 + \quad X_2 \leq 6$ (selling)

$\quad\quad\quad\quad 2X_1 + \quad X_2 \leq 9$ (mixing time)

$\quad\quad\quad\quad 2X_1 + 3X_2 \leq 16$ (shelf space)

and $\quad\quad\quad X_1, \quad X_2 \geq 0$

B.4.3 LP Problem/Model Formulation Practice: Diet Problem

Problem Statement: A diet is to contain at least 10 ounces of nutrient P, 12 ounces of nutrient R, and 20 ounces of nutrient S. These nutrients are acquired from foods

A and B. Each pound of A costs four cents and has four ounces of P, three ounces of R, and no S. Each pound of B costs seven cents and has one ounce of P, two ounces of R, and four ounces of S. Desiring minimum cost, how many pounds of each food should be purchased if the stated dietary requirements are to be met?

Formulation, by steps:

1. **Determine the type of problem**—This problem only mentions costs. Therefore, it must be a minimization problem.

2. **Define the decision variables**—How many cost values were used in Step 1 to determine the type of problem? Two (four cents and seven cents) are required. So how many decisions variables are needed? Two: If the four cents is the cost per pound of food A, the first decision variable follows:

   X_1 = number of pounds of food A to purchase

   Note that there is no time horizon (day, week, and so on) in this problem. So do not put one in. The second decision variable is as follows:

   X_2 = number of pounds of food B to purchase

3. **Formulate the objective function**—Note next that "cents" are being used. Some modelers might express the cents as 0.04 and 0.07, and others might express them as integer values. Note here that they are modeled as cents.

   Minimize: $Z = 4X_1 + 7X_2$

4. **Formulate the constraints**—This problem illustrates how the "right-hand-side strategy" for formulating constraints might be helpful. Note in the first sentence, "A diet is to contain at least 10 ounces of nutrient P, 12 ounces of nutrient R, and 20 ounces of nutrient S," how the total minimum requirements are listed. These values create a column vector (10, 12, and 20), as presented in the right-hand-side values of the constraints that follow:

$$
\begin{aligned}
\text{subject to:} \quad 4X_1 + X_2 &\geq 10 \text{ (nutrient P)} \\
3X_1 + 2X_2 &\geq 12 \text{ (nutrient R)} \\
4X_2 &\geq 20 \text{ (nutrient S)}
\end{aligned}
$$

Note how the technology coefficients for food A (the X_1 column) can be found in a single sentence, "Each pound of A costs four cents and has four ounces of P, three ounces of R, and no S," and food B (the X_2 column) can be found in a single sentence, "Each pound of B costs seven cents and has one ounce of P, two ounces of R, and four ounces of S." Because all the constraints had total minimum amounts of nutrients, the resulting expressions are all greater than or equal to.

5. **State the nonnegativity and given requirements**—Because the model has only two decision variables, all that is needed is to state the same nonnegative requirements as the basic generalized model presented in Section B.3.4:

and $X_1, X_2 \geq 0$

The entire formulation of the diet problem is again presented here:

Minimize: $Z = 4X_1 + 7X_2$

subject to:	$4X_1 + X_2$	≥ 10 (nutrient P)
	$3X_1 + 2X_2$	≥ 12 (nutrient R)
	$4X_2$	≥ 20 (nutrient S)
and	X_1, X_2	≥ 0

B.4.4 LP Problem/Model Formulation Practice: Farming Problem

Problem Statement: The Smith family owns 175 acres of farmland for breeding pigs and sheep. On average, it takes 0.5 acres of land to support either a pig or a sheep. The family can produce up to a total of 7,000 hours of labor for breeding. It takes 15 hours of labor to breed a pig and 20 hours of labor to breed a sheep. Although the family is willing to breed sheep, they do not want to breed more than 200 sheep at a time. Also, pig breeding is limited to 250. It is expected that each pig will contribute $300 profit, whereas each sheep will contribute $350.

Formulation, by steps:

1. **Determine the type of problem**—The problem only mentions profit, so it has to be a maximization problem.

2. **Define the decision variables**—The profit coefficients are attached to pigs and sheep, and there is no stated time horizon, so

X_1 = number of pigs to breed

X_2 = number of sheep to breed

3. **Formulate the objective function:**

Maximize: $Z = 300X_1 + 350X_2$

4. Formulate the constraints:

subject to:
$$0.5X_1 + 0.5X_2 \leq 175 \text{ (acres of land)}$$
$$15X_1 + 20X_2 \leq 7000 \text{ (labor hours)}$$
$$X_1 \leq 250 \text{ (maximum pig breeding)}$$
$$X_2 \leq 200 \text{ (maximum sheep breeding)}$$

5. State the nonnegativity and given requirements:

and
$$X_1, X_2 \geq 0$$

B.4.5 LP Problem/Model Formulation Practice: Customer Service Problem

Problem Statement: The customer service department of a local department store provides repair services for merchandise sold. During one week, 5 television sets, 12 radios, and 18 electric percolators were returned for repair, representing overload work items. Two repair people are temporarily employed as part-time helpers to deal with the overload work. In a normal 8-hour workday, Person 1 can repair 1 television, 3 radios, and 3 electric percolators. In a normal 8-hour workday, Person 2 repairs 1 television, 2 radios, and 6 electric percolators. Person 1 makes $55 per day, and Person 2 makes $52 per day. The customer service department wants to minimize the total cost of operation, while maintaining good customer relationships. How many days should the two repair people be employed to handle the overload of work during this one week?

Formulation, by steps:

1. **Determine the type of problem**—The problem only mentions cost, so it has to be a minimization problem.

2. **Define the decision variables**—The cost coefficients are attached to Person 1 and Person 2. Now this is a "fuzzy" time horizon problem. Are these people being hired for a week? No! They are hired for some unknown number of days to process a week's overload. So the decision variables in this problem do not need a time horizon other than to say, specifically, they are handling the overload work. It can be written as follows:

X_1 = number of days Person 1 should be hired to handle the overload work

X_2 = number of days Person 2 should be hired to handle the overload work

3. **Formulate the objective function:**

   Minimize: $Z = 55X_1 + 52X_2$

4. **Formulate the constraints:**

   subject to: $\quad X_1 + \quad X_2 \quad \geq 5$ (TVs)

   $\qquad\qquad 3X_1 + 2X_2 \quad \geq 12$ (radios)

   $\qquad\qquad 3X_1 + 6X_2 \quad \geq 18$ (percolators)

5. **State the nonnegativity and given requirements:**

   and $\qquad X_1, X_2 \geq 0$

B.4.6 LP Problem/Model Formulation Practice: Clarke Special Parts Problem

Problem Statement: The Clarke Special Parts Company manufactures three products: A, B, and C. Three manufacturing centers are necessary for the production process. Product A only passes through Centers 1 and 2; Products B and C must pass through all three manufacturing centers. The time required in each center to produce one unit of each of the three products is noted as follows:

Product	Center 1	Center 2	Center 3
A	3 hours	2 hours	0 hours
B	1	2	2
C	2	1	3

So a unit of Product A takes three hours at Center 1, two hours at Center 2, and zero hours at Center 3. Each center is on a 40-hour week. The time available for production must be decreased by the necessary cleanup time. Center 1 requires four hours of cleanup, Center 2 requires seven hours, and Center 3 requires five hours. It is estimated that the profit contribution is $60 per unit of Product A, $40 per unit of Product B, and $30 per unit of Product C. How many units of each of these special parts should the company produce to obtain the maximum profit?

Formulation, by steps:

1. **Determine the type of problem**—The problem only mentions profit, so it has to be a maximization problem.

2. **Define the decision variables**—The profit coefficients are attached to Products A, B, and C and have a weekly stated time horizon, so

X_1 = number of units of Product A to produce per week

X_2 = number of units of Product B to produce per week

X_3 = number of units of Product C to produce per week

3. **Formulate the objective function:**

Maximize: $Z = 60X_1 + 40X_2 + 30X_3$

4. **Formulate the constraints**—This problem illustrates that some arithmetic may be needed to derive model parameters. In this case, the right-hand-side b values need to be adjusted for the cleanup time. In a week, each department starts with 40 hours for production purposes. They then have to be decreased for the cleanup time, as stated in the problem sentences, "The time available for production must be decreased by the necessary cleanup time. Center 1 requires four hours of cleanup, Center 2 requires seven hours, and Center 3 requires five hours." So, for Center 1 we have 36 hours (40 – 4), for Center 2 we have 33 hours (40 – 7), and so on to formulate the right-hand-side values in each constraint. This problem also illustrates the use of the left-hand-side strategy for formulating constraints. Note how the tabled values are the technology coefficients listed by columns in the constraints that follow:

subject to:
$$3X_1 + X_2 + 2X_3 \leq 36 \text{ (Center 1)}$$
$$2X_1 + 2X_2 + X_3 \leq 33 \text{ (Center 2)}$$
$$2X_2 + 3X_3 \leq 35 \text{ (Center 3)}$$

5. **State the nonnegativity and given requirements:**

and $X_1, X_2, X_3 \geq 0$

B.4.7 LP Problem/Model Formulation Practice: Federal Division Problem

Problem Statement: The Federal Division has a contract to supply at least 72 engine parts. There are three different production processes for engine parts. The processes require different amounts of skilled labor, unskilled labor, and computer time for machine tools. Any one process is, by itself, capable of producing an engine part.

Process	Hours of Skilled Labor for One Engine Part	Hours of Unskilled Labor for One Engine Part	Computer Time in Minutes for One Engine
1	3	4	1
2	6	4	0
3	0	1	4

In the foreign country where it operates its plant, skilled labor costs $8/hour, and no more than 288 hours can be obtained. Unskilled labor costs $3/hour, and no more than 324 hours can be obtained. Computer time costs $10/minute, and no more than 196 minutes are available. Recommend a course of action.

Formulation, by steps:

1. **Determine the type of problem**—The problem only mentions cost, so it has to be a minimization problem.

2. **Define the decision variables**—This problem is meant to challenge and build skills in identifying decision variables. What is the variable here? In this problem, only one product, an engine part, is produced. So what is the variable? The variable in this problem is the "process" by which engine parts are produced. So the decision variables become this:

X_1 = number of engine parts to produce by Process 1

X_2 = number of engine parts to produce by Process 2

X_3 = number of engine parts to produce by Process 3

Like many firms today that have older technology to produce current products, this problem seeks to make the best use of a combination of old and new process technologies to produce the single product called engine parts.

3. **Formulate the objective function**—Given the definition of the preceding decision variables as processes, identify the correct contribution coefficients. Because they are directly related to the decision variable definitions, they can be defined as follows:

c_1 = cost of producing one engine part by Process 1

c_2 = cost of producing one engine part by Process 2

c_3 = cost of producing one engine part by Process 3

How are these parameters found? Use a little arithmetic to compute them as follows:

Cost of Skilled Labor	Cost of Unskilled Labor	Cost of Computer Time
c_1 = \$8 × 3 hours +	\$3 × 4 hours +	\$10 × 1 minute = \$46
c_2 = \$8 × 6 hours +	\$3 × 4 hours +	\$10 × 0 minutes = \$60
c_3 = \$8 × 0 hours +	\$3 × 1 hour +	\$10 × 4 minutes = \$43

The resulting three parameters can then be put in an objective function as follows:

Minimize: $Z = 46X_1 + 60X_2 + 43X_3$

4. **Formulate the constraints:**

subject to:
$$X_1 + X_2 + X_3 \geq 72 \text{ (Supply)}$$
$$3X_1 + 6X_2 \leq 288 \text{ (Skilled labor)}$$
$$4X_1 + 4X_2 + X_3 \leq 324 \text{ (Unskilled labor)}$$
$$X_1 + 4X_3 \leq 196 \text{ (Computer time)}$$

5. **State the nonnegativity and given requirements:**

and $X_1, X_2, X_3 \geq 0$

There are additional practice problems in Section B.8.

B.5 Computer-Based Solutions for Linear Programming Using the Simplex Method

In this section, we examine computer-based solutions methods. The most common method to obtain a solution for an LP model is through the use of the simplex algorithmic method. Although some elements of this methodology are useful in understanding LP solutions, our focus will be on utilizing computer software to generate answers.

B.5.1 Introduction

The simplex method is an algebraic methodology based on finite mathematics. Remember determinates or matrix algebra from high school? The simplex method is based on the same mathematical process. The computer will generate a solution using the simplex method, although it is not needed to know how the mathematical process works. What is important is to understand that the simplex method is an optimization process. So it not only gives an optimal solution, but it internally proves that the solution is optimal. This section seeks to provide additional understanding of the by-products of information that the simplex method's solution provides. These business analytics are often viewed as important as the solution that the LP model is designed to generate.

B.5.2 Simplex Variables

The simplex method determines the optimal values for the X_j decision variables and the value of Z. Using the simplex method requires employing three other variables as well:

- **Slack variable**—A slack variable is used in a less than or equal to constraint to permit the left-hand-side of the constraint to equal the right-hand-side in the beginning of the solution process. It works like this. Given the following constraint:

$$X_1 + X_2 \leq 100$$

if one wants to express it as an equality, one would have to add an additional variable to take up the slack if the sum of the product is less than the 100 of the right-hand-side value. The slack variable is added to the left-hand-side of the constraint and rewritten as an equality expression:

$$X_1 + X_2 + s_1 = 100$$

For each constraint that is modeled, add a different slack variable. For example, in the farming problem from Section B.4.4, the four constraints can be expressed as simplex equality constraints:

$$0.5X_1 + 0.5X_2 + s_1 \qquad\qquad = 175 \text{ (Acres of land)}$$
$$15X_1 + 20X_2 \qquad + s_2 \qquad = 7000 \text{ (Labor hours)}$$
$$X_1 \qquad\qquad\qquad + s_3 \qquad = 250 \text{ (Maximum pig breeding)}$$
$$X_2 \qquad\qquad + s_4 = 200 \text{ (Maximum sheep breeding)}$$

Why does the simplex method require the constraints to be expressed as equalities? In an optimal solution, one might not need to use all the maximum resources (acres of land, labor hours, and so on). If they're not needed, they become slack resources. As it turns out, the slack variables become as important for managerial decision-making as the decision variables, because slack resources are idle resources that can be reallocated to more profitable production activities.

- **Surplus variable**—A surplus variable is used in a greater than or equal to constraint to permit the left-hand-side of the constraint to equal the right-hand-side in the beginning of the solution process. Given the following constraint:

$$X_1 + X_2 \geq 100$$

if one wants to express it as an equality, an additional variable would have to be added to take up the additional or surplus if the sum of the product is greater than the 100 of the right-hand-side value. The surplus variable is added to the right-hand-side of the constraint and rewritten like this:

$$X_1 + X_2 = 100 + s_1$$

We then have to subtract the surplus variable from both sides to put it on the left-hand-side, where all variables belong (a constraint must have all the variables on the left-hand-side and a constant "b" value on the right-hand-side to be a valid constraint). Note the following expression:

$$X_1 + X_2 - s_1 = 100$$

Unfortunately, the negative sign in front of a variable (even a surplus variable) is not acceptable in the mathematical process of the simplex method. So still another variable will have to be created to temporarily cancel out the negativity of the surplus variable in the simplex process. This third new variable is called an artificial variable and is represented by a capital "A" in the expression that follows:

$$X_1 + X_2 - s_1 + A_1 = 100$$

- **Artificial variable**—The sole purpose of the artificial variable is just to perform a temporary mathematical adjustment to permit the simplex process to handle the negativity of the surplus variable. Ideally, the artificial variable will never pop up in a model solution. (There will be an LP complication that we will discuss later, where the artificial variable can pop up and prevent an optimal solution from happening.)

In summary, the slack and surplus variables are not only necessary for the simplex method to work, but they provide useful information on a resulting solution by explaining deviation from right-hand-side values and revealing excess resources for reallocation.

B.5.3 Using the LINGO Software for Linear Programming Analysis

There are many software applications that can solve LP problems. In this section, we will examine the software app LINGO.

B.5.3.1 Trial Versions of LINGO Software (as of January 2014)

LINGO software is a product of Lindo Systems (www.lindo.com). The use of this software can be made available for a limited time as a demo for free. For purposes of this book, the limited time will be sufficient. For those interested in owning a copy, there is an inexpensive version available through the Lindo website. Microsoft® Windows® versions of LINGO are compatible with Windows 2000, Windows XP, Windows Vista, Windows 7, and Windows 8. To obtain the trial version (useful for this book), complete the following steps:

1. Go to www.lindo.com.
2. Click on the LINGO icon.
3. Click on Download a Trial Version.
4. Click on Download LINGO.
5. Click on the appropriate download for the LINGO version that best works on your computer system.
6. The system may request that you register your copy. Feel free to do it now or later.

To confirm that you downloaded it correctly, enter the LP model using the explanation in Section B.5.3.2.

B.5.3.2 How to Use LINGO to Generate a Solution

To use the LINGO software, which incorporates the simplex method, it is necessary to enter the data from the LP problem/model formulation into the computer. The input process and solution interpretation are illustrated using the farming problem from Section B.4.4, as stated again here:

Maximize: $Z = 300X_1 + 350X_2$

subject to: $0.5X_1 + 0.5X_2 \leq 175$ (acres of land)

$15X_1 + 20X_2 \leq 7000$ (labor hours)

$X_1 \leq 250$ (maximum pig breeding)

$X_2 \leq 200$ (maximum sheep breeding)

and $X_1, X_2 \geq 0$

where:

X_1 = number of pigs to breed

X_2 = number of sheep to breed

Use LINGO by simply entering a modified version of the LP model formulation employing the following steps:

1. Double-click on the LINGO icon on the desktop (or wherever it's located on the computer). A blank window opens. This is where to enter the LP model formulation.

2. The farming problem/model formulation should be entered, as stated in Figure B.1.

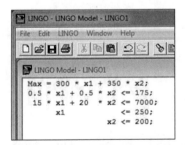

Figure B.1 Farming problem input into LINGO

Note: (1) Use the terms Max or Min. (2) Do not use the Z parameter in the model formulation. (3) Use an asterisk between parameters and variables. (4) Variables may be whole words or a combination of letters or numbers, but they must not contain spaces or special characters. (5) Do not state "subject to." (6) Each expression must end in a semicolon (;). (7) Do not state given requirements. (8) The less than or equal to symbol is expressed as $\leq$.

3. Click on the LINGO menu option at the top of the window to reveal the SOLVE option. Click it.

4. If anything is incorrectly input, an error statement in the form of a LINGO ERROR MESSAGE window pops up and shows you where the first mistake was made. Recheck the input data, just as it is presented in Figure B.1.

5. Assuming everything is correctly entered and the SOLVE option has been clicked, two windows will pop up. The first window, LINGO SOLVER STATUS, is a summary window to provide details on the interactive process of the simplex method. This is not essential information. Simply click CLOSE and exit the window. The second window presents the solution to the problem. The solution for the farming problem is presented in Figure B.2.

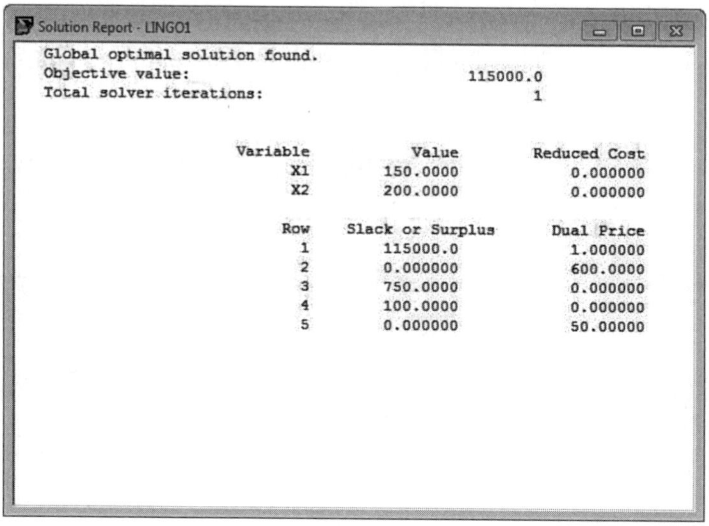

```
Solution Report - LINGO1                                         _  □  ☒

  Global optimal solution found.
  Objective value:                        115000.0
  Total solver iterations:                     1


              Variable          Value         Reduced Cost
                 X1           150.0000           0.000000
                 X2           200.0000           0.000000

              Row       Slack or Surplus       Dual Price
               1            115000.0            1.000000
               2            0.000000            600.0000
               3            750.0000            0.000000
               4            100.0000            0.000000
               5            0.000000            50.00000
```

Figure B.2 Farming problem solution by LINGO

The SOLUTION REPORT provides the solution for the LP problem. This report has several parts, some of which are explained in Appendix C, "Duality and Sensitivity Analysis in Linear Programming." For now, focus on just reading the solution to the farming problem. In the OBJECTIVE FUNCTION row, the value of 115000.0 is the optimal value for Z, which in this problem is $115,000. The TOTAL SOLVER ITERATIONS row is just a notification that it took one iteration of the simplex method to generate the solution (not important information at the present). The optimal decision variable values are provided in the column headings listed as VARIABLE and VALUE. The REDUCED COST and DUAL PRICE columns can be ignored for now and will be discussed in Appendix C. Each x_j decision variable is listed by row in these columns, and each has its optimal value given in the VALUE column. So, for this farming problem, X_1 = 150 and X_2 = 200, which means one should breed 150 pigs and 200 sheep to achieve the maximize profit of \$115,000.

In addition to the optimal Z and decision variable values, the simplex method gives the optimal slack and surplus variable values. This problem only had less than or equal to constraints, so there are only slack variables in each constraint. The optimal slack values are given by row. The rows (1 to 5) are a listing of the five expressions, including the objective function. Ignoring Row 1 for now, the four constraints are identified by Rows 2, 3, 4, and 5. The optimal slack values for each constraint are found in the SLACK AND SURPLUS column. So, in this problem, S_1 = 0, S_2 = 750, S_3 = 100, and S_4 = 0. These numbers can be checked by substituting the optimal decision variable values into each constraint as stated here:

$$0.5(150) + 0.5(200) + S_1 = 175 \text{ (Acres of land)}$$
$$15(150) \ + 20(200) \ + S_2 = 7000 \text{ (Labor hours)}$$
$$(150) \qquad\qquad\quad + S_3 = 250 \text{ (Maximum pig breeding)}$$
$$(200) \qquad\qquad\quad + S_4 = 200 \text{ (Maximum sheep breeding)}$$

It will always be true that the slack variables will equal the values given in the SLACK AND SURPLUS column to permit the equalities to hold true:

$$175 \ + 0 \quad = 175 \text{ (Acres of land)}$$
$$6250 + 750 = 7000 \text{ (Labor hours)}$$
$$150 \ + 100 = 250 \text{ (Maximum pig breeding)}$$
$$200 \ + 0 \quad = 200 \text{ (Maximum sheep breeding)}$$

The fact that 750 hours of labor are not needed is valuable information. One can now reallocate those hours to some other farming activity not specified in the original problem statement. Also, in the first column of numbers, the resulting optimal usage of the right-hand-side values can be seen. These values are obtained by subtracting from the b parameters in the stated formulation of the LP model the resulting slack variable values. In this solution, the actual usage of the b parameters follows:

$$b_1 = 175 \text{ (i.e., } 175 - 0)$$
$$b_2 = 6250 \text{ (i.e., } 7000 - 750)$$
$$b_3 = 150 \text{ (i.e., } 250 - 100)$$
$$b_4 = 200 \text{ (i.e., } 200 - 0)$$

Knowing how many units of resources will be used in an optimal production or farming problem is just as important for planning purposes as knowing how many units of product or animal one plans to produce.

Here's a question that comes up: How does one know that the values in the Slack/Surplus column are slack values and not surplus values? Check the direction of the inequalities in the input section of the printout to know for sure. If the direction of the inequalities is less than or equal to, the values in the Slack/Surplus column have to be slack values. If the inequalities are greater than or equal to, the values have to be surplus. In some problems, these business analytics are what is most important in planning resources.

The existence of slack or surplus also reveals which constraints are necessary for a given solution and which constraints are not. A constraint that has zero slack or surplus is called a *binding constraint* because the constraint is important in determining the solution. Specifically, this constraint's resources directly constrained the decision variables values. A constraint that has a positive slack or surplus value is called a *nonbinding constraint* or *redundant constraint* because it does not impact the solution in any way. In fact, nonbinding constraints can be dropped from a model, and one will find the values of the decision variables will be the same. In the farming problem, the first and fourth constraints are binding, and the second and third are nonbinding.

Now consider another computer-generated solution for a different problem, such as the minimization problem presented here:

Minimize $Z = 20X_1 + 2X_2$

$$
\begin{aligned}
\text{subject to:} \quad 2X_1 + X_2 &\geq 10 \\
X_1 + X_2 &\geq 5 \\
5X_1 + 10X_2 &\geq 50 \\
\text{and} \qquad\qquad X_1, X_2 &\geq 0
\end{aligned}
$$

In Figure B.3, the LINGO model input is presented. The LINGO solution is shown in Figure B.4.

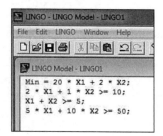

Figure B.3 LINGO minimization model input

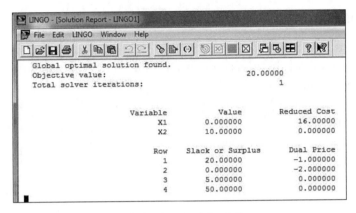

Figure B.4 LINGO minimization model solution

In Figure B.4, the resulting optimal values for Z and the decision variables are $Z = 20$, $X_1 = 0$, and $X_2 = 10$. The values of the surplus variables in this problem are $S_1 = 0$, $S_2 = 5$, and $S_3 = 50$.

Note that these are surplus variables, because the constraints are all greater than or equal to. The values of the actual b parameters used in the solution are obtained by adding the stated b values from the model formulation and the related surplus variable values. The resulting actual values of b that will be utilized by this solution are $b_1 = 10$ (10 + 0), $b_2 = 10$ (5 + 5), and $b_3 = 100$ (50 + 50).

There are additional computer-generated practice problems in Section B.8.

The simplex method is a powerful analytic methodology for obtaining solutions from LP problems. Unfortunately, complications can develop that prevent users from obtaining a solution.

B.6 Linear Programming Complications

There are complications that prevent the simplex method from generating a desired optimal solution or even the ability to formulate a problem. Being aware of these complications and what causes them can help users overcome them. Some of these complications include unbounded solutions, infeasible solutions, blending formulations, and multidimensional variables.

B.6.1 Unbounded Solutions

An unbounded solution is not, in fact, a solution. The formulation of the problem is incorrect such that one or more of the decision variable values in the model goes to positive infinity. The resolution of this complication is to reformulate the problem correctly. How does one know the problem is unbounded? Most software packages tell the user when they run the problem. An example of an unbounded problem expressed as LINGO input is presented in Figure B.5, and its nonsolution is presented in Figure B.6.

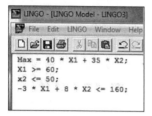

Figure B.5 Unbounded LINGO maximization model input

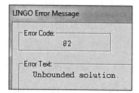

Figure B.6 Unbounded LINGO maximization model nonsolution notification

B.6.2 Infeasible Solutions

An infeasible solution is not a solution. The model has been incorrectly formulated in such a way that no solution set could be found to satisfy all the constraints in the model. Unless a solution is at least feasible, it cannot possibly be optimal. The resolution of this complication is to reformulate the problem correctly. How does one know the problem is infeasible? Most software packages will tell the user when they run the problem. An example of an infeasible problem expressed as LINGO input is presented in Figure B.7, and its nonsolution is presented in Figure B.8.

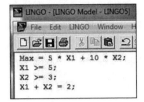

Figure B.7 Infeasible LINGO maximization model input

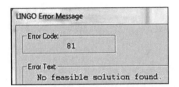

Figure B.8 Infeasible LINGO maximization model nonsolution notification

B.6.3 Blending Formulations

A *blending formulation* is a formulation complication. In some situations, there may be a need to achieve a blend of two or more items, like mixing ounces of cereal with ounces of fruit to make a new breakfast product.

This is accomplished in LP models with constraints. A simple example is a one-to-one ratio between two decision variables. For example, suppose there are several decision variables in a model (X_2 and X_3), but it is desired to have them equal each other in the final optimal solution. How can this relationship be expressed as a constraint to achieve this ratio? Simply, one wants $X_2 = X_3$, which is not a constraint with a constant b right-hand-side value. So algebraically it is converted to $X_2 - X_3 = 0$ or $-X_2 + X_3 = 0$. Either of these two equalities achieves a one-to-one equality for the two decision variables in a solution for an LP model.

Another more complex blending formulation might involve mixing unequal parts. For example, suppose one wants to mix two parts of X_1 to every one part of X_2. The means of formulation of this constraint is achieved by a simple step-wise ratio approach between the two parts:

1. Express the two mixture ratios (two parts of X_1 to one part of X_2) as equalities:

$$2 = X_1$$
$$1 = X_2$$

2. Then set them as ratios:

$$\frac{2}{1} = \frac{X_1}{X_2}$$

3. Then algebraically multiply each side to obtain the equation:

$$2X_2 = 1X_1$$

4. Finally, subtract $2X_2$ from both sides to obtain the desired constraint:

$$1X_1 - 2X_2 = 0$$

This approach can be used to develop as many blending constraints as needed to achieve a desired mixture. It is important to remember, though, that these ratio constraints must be on a one-variable to one-variable basis. One limitation of this process is that there cannot be more than two variables in a constraint at a time, but there can be multiple blending of constraints in a single model.

B.6.4 Multidimensional Decision Variable Formulations

A multidimensional decision variable is one that has more than one characteristic describing it in its definition. An example of how one might structure and use multidimensional decision variables can be seen in a typical human resource problem. Suppose there are two people who must determine the optimal number of hours they should be scheduled to work over the next seven days. The configuration of this problem can best be expressed in a two-dimensional layout below:

			Days of the Week				
Person	**Monday**	**Tuesday**	**Wednesday**	**Thursday**	**Friday**	**Saturday**	**Sunday**
1	X_{11}	X_{12}	X_{13}	X_{14}	X_{15}	X_{16}	X_{17}
2	X_{21}	X_{22}	X_{23}	X_{24}	X_{25}	X_{26}	X_{27}

So these variables have two dimensions: a person dimension and a day dimension. The variables are always expressed with the first subscript being the row (the person) and the second subscript being the column (day of the week). These variables could be generalized as X_{ij} = number of hours the ith person should work on the jth day.

This type of decision variable permits one to structure constraints in two dimensions. For example, suppose one has to limit the number of hours Person 1 could work in the week to no more than 60. The resulting constraint would be used:

$$X_{11} + X_{12} + X_{13} + X_{14} + X_{15} + X_{16} + X_{17} \geq 60$$

Now suppose we also want to limit the number of hours for both employees on a Saturday to no more than 10. This day type of constraint would be as follows:

$X_{16} + X_{26} \geq 10$

Hence, the solution procedure would seek a value for these variables in both a "person" dimension and a "day" dimension.

The number of dimensions used for decision variables is up to the modeler. Remember, each dimension brings with it an almost geometric increase in the number of decision variables in the model.

B.7 Necessary Assumptions for Linear Programming Models

Five basic assumptions must be met for LP to be used in modeling a situation. These assumptions are also useful in deciding whether LP should be used to model a problem. Here are the five assumptions:

1. **Linearity**—All constraints and the objective function must be linear. If one has a nonlinear profit or cost function or a nonlinear constraint, other nonlinear programming methodologies must be used. A number of these are available in the literature, including such techniques as quadratic programming, separable programming, and Kuhn-Tucker conditions.

2. **Additivity**—All the constraints and the objective function must for any value of the decision variables add up exactly as modeled. That is, one cannot have synergistic impact, where 2 + 2 = 5. Regardless of the size of the decision variable value, the added values of the coefficients must be the simple sum of the products. If they're not, use another methodology.

3. **Divisibility**—In the LP models presented in this appendix, the nonnegativity and given requirements allow the decision variable values to be real numbers or any fractional value. This means that if a decision variable ended up being 0.5, one-half of the profit or cost of that decision variable is exactly what will be received. Also, if the labor hour usage of that decision variable is two, then 0.5 of two means that exactly one hour of labor will be used. Sometimes fractional answers are not realistic. In such cases, use something other than LP—perhaps integer programming (which will be discussed in Appendix D).

4. **Finiteness**—This requirement simply means that the values of the decision variables must be finite. If they are not finite, they are infinite and, therefore, unbounded.

5. **Certainty and a static time period**—All of the a, b, and c parameters of an LP model must be known with certainty. We can help ensure this certainty by stating a time horizon or a static time period when the decision variables are defined. The static time period specifies the period over which the answer and the parameters remain true.

B.8 Linear Programming Practice Problems

Following are several practice LP problems, followed by their answers. Use these problems to practice the methodologies and concepts presented in this appendix.

1. A small furniture manufacturer produces three different kinds of furniture: desks, chairs, and bookcases. The wooden materials have to be cut properly by machines. In total, 100 machine hours are available for cutting. Each unit of desks, chairs, and bookcases requires 0.8 machine hours, 0.4 machine hours, and 0.5 machine hours, respectively. This manufacturer also has 650 labor hours available for painting and polishing. Each unit of desks, chairs, and bookcases requires 5 labor hours, 3 labor hours, and 3 labor hours for painting and polishing, respectively. These products are to be stored in a warehouse, which has a total capacity of 1,260 sq. ft. The floor space required by these three products is 9 sq. ft., 6 sq. ft., and 9 sq. ft., respectively, per unit of each product. In the market, each product is sold at a profit of $30, $16, and $25 per unit, respectively. What is the formulation of this problem to determine how many units of each product should be made to realize a maximum profit? Answer: Let X_1, X_2, and X_3 be the number of units of desks, chairs, and bookcases to be produced, respectively. Because 100 total machine hours are available for cutting, the production of X_1, X_2, and X_3 should utilize no more than the available machine hours. Therefore, the mathematical statement of the first constraint is in the form $0.8X_1 + 0.4X_2 + 0.5X_3 \leq 100$. Also, no more than 650 labor hours and 1,260 sq. ft. are available for painting, polishing, and storing, respectively. Therefore, these two constraints are in the form $5X_1 + 3X_2 + 3X_3 \leq 650$ and $9X_1 + 6X_2 + 9X_3 \leq 1,260$. Finally, the decision variables must be nonnegative. The complete problem formulation follows:

Maximize: $Z = 30X_1 + 16X_2 + 25X_3$

subject to:

$$0.8X_1 + 0.4X_2 + 0.5X_3 \leq 100$$
$$5X_1 + 3X_2 + 3X_3 \leq 650$$
$$9X_1 + 6X_2 + 9X_3 \leq 1{,}260$$

and

$$X_1, X_2, X_3 \geq 0$$

The LINGO input data and solution are presented in Figures B.9 and B.10.

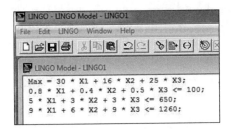

Figure B.9 LINGO practice problem 1 model input

```
Solution Report - LINGO1

Global optimal solution found.
Objective value:                          4000.000
Total solver iterations:                         0

            Variable           Value       Reduced Cost
                  X1        100.0000           0.000000
                  X2        0.000000           1.777778
                  X3        40.00000           0.000000

                 Row   Slack or Surplus        Dual Price
                   1          4000.000          1.000000
                   2          0.000000         16.66667
                   3          30.00000          0.000000
                   4          0.000000          1.851852
```

Figure B.10 LINGO practice problem 1 solution

2. The Riverside Company wants to outsource production of three products: premium toys, deluxe toys, and regular toys. These three different toys can be produced at two different external plants with different production capacities. In a normal day, Outsource Plant A produces 20 premium toys, 30 deluxe toys, and 100 regular toys. Outsource Plant B produces 50 premium toys, 40 deluxe toys, and 60 regular toys. The monthly demand for each is known to be 4,000 units, 3,000 units, and 1,000 units, respectively. The company has to pay a daily cost of operation, which is $50,000 for Outsource Plant A and $40,000 for Outsource

Plant B. What is the formulation of this problem to find the optimum number of days of operation per month at the two different plants to minimize the total cost while meeting the demand? Answer: Let the decision variables X_1 and X_2 represent the number of days of operation in each of the plants. The objective function of this problem is the sum of the daily operational costs in the two different plants expressed as Minimize: $Z = 50,000X_1 + 40,000X_2$. The objective is to determine the value of the decision variables, X_1 and X_2, which yields the minimum total cost subject to the constraints. The production of each of the three types of toys must be at least greater than or equal to the specific quantity to meet demand requirements. In no event should the production be less than the quantities of each product demanded. Together with the constraints, the problem can be formulated as such:

Minimize: $Z = 50,000X_1 + 40,000X_2$

subject to:
$$20X_1 + 50X_2 \leq 4,000$$
$$30X_1 + 40X_2 \leq 3,000$$
$$100X_1 + 60X_2 \leq 1,000$$

and
$$X_1, X_2 \geq 0$$

The LINGO input data and solution are presented in Figures B.11 and B.12.

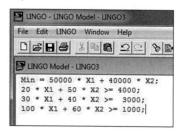

Figure B.11 LINGO practice problem 2 model input

```
Solution Report - LINGO3
Global optimal solution found.
Objective value:                    3200000.
Total solver iterations:                   1


           Variable          Value        Reduced Cost
               X1         0.000000           34000.00
               X2         80.00000          0.000000

           Row      Slack or Surplus       Dual Price
             1         3200000.           -1.000000
             2         0.000000            -800.0000
             3         200.0000            0.000000
             4         3800.000            0.000000
```

Figure B.12 LINGO practice problem 2 solution

3. The Turned-On Radio Company manufactures models A, B, and C, which have profit contributions of 8, 15, and 25, respectively, per unit. The weekly minimum production requirements are 100 for model A, 15 for model B, and 75 for model C. Each type of radio requires a certain amount of time for the manufacturing component parts, assembling, and packaging. Specifically, a dozen units of model A require three hours of manufacturing, four hours for assembling, and one hour for packaging. The corresponding figures for a dozen units of model B are 3.5, 5, and 1.5; for a dozen units of model C are 5, 8, and 3. During the forthcoming week, the company has available 150 hours of manufacturing, 200 hours of assembling, and 60 hours of packing time. What is the formulation of the production scheduling problem as a linear programming model? Answer: Let X_1 = the number of units of model A to produce, X_2 = the number of units of model B to produce, and X_3 = the number of units of model C to produce. The formulation of this problem in an LP model is given here:

Maximize: $Z = 8X_1 + 15X_2 + 25X_3$

subject to:

$X_1 \geq 100$ (Min. Production)

$X_2 \geq 15$ (Min. Production)

$X_3 \geq 75$ (Min. Production)

$0.25X_1 + 0.29X_2 + 0.42X_3 \geq 150$ (Manufacturing)

$0.33X_1 + 0.42X_2 + 0.67X_3 \geq 200$ (Assembly)

$0.08X_1 + 0.13X_2 + 0.25X_3 \geq 60$ (Packaging)

and $X_1, X_2, X_3 \geq 0$

The LINGO input data and solution are presented in Figures B.13 and B.14. Note that the fractional values have been converted to decimal values because the LINGO software requires no special characters with a parameter, such as a backslash.

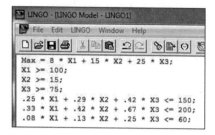

Figure B.13 LINGO practice problem 3 model input

```
LINGO - [Solution Report - LINGO1]
File   Edit   LINGO   Window   Help

 Global optimal solution found.
 Objective value:                            6511.538
 Total solver iterations:                           3

                Variable           Value        Reduced Cost
                      X1        100.0000            0.000000
                      X2        255.7692            0.000000
                      X3        75.00000            0.000000

                     Row   Slack or Surplus          Dual Price
                       1          6511.538            1.000000
                       2          0.000000           -1.230769
                       3          240.7692            0.000000
                       4          0.000000           -3.846154
                       5          19.32692            0.000000
                       6          9.326923            0.000000
                       7          0.000000          115.3846
```

Figure B.14 LINGO practice problem 3 solution

C

Duality and Sensitivity Analysis in Linear Programming

C.1 Introduction

This appendix is a continuation of the subject of linear programming (LP). The topics covered in this appendix are duality and sensitivity analysis in linear programing models. These methods are of value in the third step of the BA process: *prescriptive analytics*.

C.2 What Is Duality?

Solving an LP problem as in Appendix B, "Linear Programming," involves solving for the optimal value for Z, X_j, and s_i. This is known as a *primal problem,* because an optimal solution is being sought that explains the primary or primal relationship between the X_j and s_i variables, values as they consume right side b_i constant parameters. Embedded within this primal solution is a *dual solution*. Every primal problem has a *dual problem* and a *dual solution*. The dual solution provides economic trade-off information (usually expressed in dollars) on the value of the right side b_i unit usages of resources. For example, suppose one solves an LP model with a Z equal to $10,000 and uses 1,000 hours of skilled labor. Among other things, the dual solution would indicate what the dollar value of each hour of skilled labor is contributing to Z, while considering the other used resources.

Duality in LP is a means of determining the economic value of right side b_i values. It is a by-product of the simplex method and is usually given in the printout of any LP simplex model solution. The focus in this appendix is to know where the dual solution is on the computer printout and what it means for business analytic analysis and decision-making purposes.

C.2.1 The Informational Value of Duality

What is the value of the information that duality provides? Duality has been used for many purposes. One important application is using duality to determine the economic contribution of resources. That is, determine the dollar contribution each resource is responsible for in generating the total optimized Z value. Another application that has appeared in the literature is its use in cost accounting, where dollar costs can be attributed to the specific resources that are used in the LP model solution.

The dual solution is an economic valuation methodology that provides valuations based on the scarcity of resources used in constraining an existing optimal solution. So, when resources are abundant (a slack resource or a surplus resource), the dual solution assigns a $0 contribution to that resource, even though some of it may be used in an optimal solution.

In summary, there are two types of information defined in the dual solution: (1) marginal contribution to Z for each unit of the related right side parameters, b_i and (2) marginal cost or loss to Z for each unit of the related decision variables, X_j. In this appendix, the focus is on using the information from the dual solution, rather than formulation or computational effort.

C.2.2 Sensitivity Analysis

Dual solutions have limitations. Specifically, the value of a dual variable only remains true for a certain number of units of the related right side value. For example, if in the farming problem (see Appendix B, Section B.4.4) one wanted to increase acres of land from the current level of 175 to 500, one cannot expect to receive a dual decision variable value of $600 for all of them. The question then becomes how many units can be increased or decreased from a given right side value and still be sure of receiving the dual decision variable value exactly. The answer to this range of numbers is found in a subsequent procedure called sensitivity analysis. *Sensitivity analysis* is a procedure to determine the limitations on parameter changes in a model and their impact on an existing solution. Although there are many different types of sensitivity analyses, the discussion will be limited to two of the most important that involve changes in the parameters c_j and b_i. In general, sensitivity analysis can answer questions related to making changes in the parameters (c_j and b_i), inform what impact those changes will have on the value of Z, and determine whether the variables in the primal solution will change.

Both c_j and b_i sensitivity analyses result in generating a *relevant range* (computed values that define the interval of allowable change in a parameter) over which changes in the two c_j and b_i parameters can take place and predict changes in the existing solution without rerunning the model. Why not just make a change in a parameter and simulate the outcomes in the model to find out the changes? That deterministic simulation approach can work in some situations, but finding specific values in which a dramatic change in an existing solution will take place (a particular threshold) could take an infinite number of experiments to find the exact threshold value. Sensitivity analysis gives the exact value and does so as a by-product of the original solution effort in solving the primal LP problem.

C.3 Duality and Sensitivity Analysis Problems

Consider four problems: two primal maximization problems and two primal minimization problems.

C.3.1 A Primal Maximization Problem

Take the farming problem formulated in Appendix B, Section B.4.4, which was solved using LINGO in Section B.5.3 (restated here):

Problem Statement: The Smith family owns 175 acres of farmland for breeding pigs and sheep. On average, it takes 0.5 acres of land to support either a pig or a sheep. The family can produce up to 7,000 hours of labor for breeding. It takes 15 hours of labor to breed a pig and 20 hours of labor to breed a sheep. Although the family is willing to breed sheep, they do not wish to breed more than 200 sheep at one time. Also, breeding pigs is limited to 250. It is expected that each pig will contribute $300 to profit, whereas each sheep will contribute $350.

$$\text{Maximize: } Z = 300X_1 + 350X_2$$

subject to:

$$0.5X_1 + 0.5X_2 \leq 175 \text{ (Acres of land)}$$
$$15X_1 + 20X_2 \leq 7000 \text{ (Labor hours)}$$
$$X_1 \leq 250 \text{ (Maximum pig breeding)}$$
$$X_2 \leq 200 \text{ (Maximum sheep breeding)}$$

and

$$X_1, X_2 \geq 0$$

where:

X_1 = number of pigs to breed

X_2 = number of sheep to breed

The LINGO formulation and data entry are presented in Figure C.1.

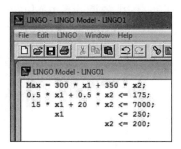

Figure C.1 Formulation of farming problem in LINGO

The dual values related to each of the four right side values b_i that one seeks to determine follow:

- Marginal contribution of one acre of land
- Marginal contribution of one hour of labor
- Marginal contribution of one pig
- Marginal contribution of one sheep

The dual values are related to each of the primal problem's X_j decision variables:

- Marginal cost or loss of one pig
- Marginal cost or loss of one sheep

It is absolutely necessary to fully understand the components of the primal problem to define the dual variables. Use either the LINGO printout of this solution presented in Figure C.2.

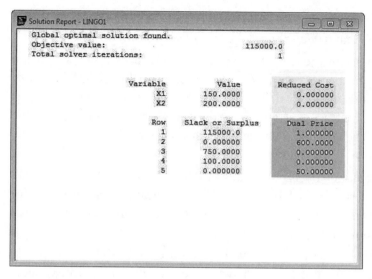

Figure C.2 LINGO LP solution to the farming problem with dual solution

The Reduced Cost column in Figure C.2 lists the dual values, and the Dual Price column lists the dual right side values. The constraints are listed by number for reference. To make the comparison, the primal and related dual solution values are detailed here:

Primal Solution

Values	(Slack)	Related Dual Solution Values
$b_1 = 175$	$(s_1 = 0)$	600 (marginal contribution of an acre of land)
$b_2 = 7000$	$(s_2 = 750)$	0 (marginal contribution of an hour of labor)
$b_3 = 250$	$(s_3 = 100)$	0 (marginal contribution of a pig)
$b_4 = 200$	$(s_3 = 0)$	50 (marginal contribution of a sheep)
$X_1 = 150$		0 (marginal cost or loss of a pig)
$X_2 = 200$		0 (marginal cost or loss of a sheep)

Note that the slack variables for the primal solution are included. These are important in understanding the resulting values of the dual problem. A few points can be made with this first solution:

- Whenever the primal solution has a positive slack value ($s_2 = 750$ and $s_3 = 100$) (or surplus), the related dual price decision variable (0 and 0) is expected to be zero. This is because in economics, scarcity of resources means that one has more than needed (slack resources). So there is no economic value in obtaining more of a resource that will just be relegated to slack resources.

- In most cases, whenever the primal solution has a zero slack value ($s_1 = 0$ and $s_4 = 0$) (or surplus in the case of minimization problem), the related dual price decision variable (600 and 50) will be positive. This is so because the zero slack (or surplus) means that the constraint is binding and has a direct impact on the existing solution. The amount of the impact per unit of decision variable value is given by the dual decision variable values.

- When the primal solution decision variables are positive ($x_1 = 150$ and $x_2 = 200$), the dual reduced cost value will be zero (0 and 0), indicating there is no cost or loss in producing the amount of the decision variable values suggested in the primal problem solution.

- When the primal solution decision variables are zero (which did not happen in the problem), the dual reduced cost values will always be some positive value, indicating that there is a cost or loss in producing each unit of the related decision variable in the primal problem.

In the farming problem, the dual price value of 600 means that the marginal contribution of one acre of land is $600. So, what is a good dual price to pay to buy an additional acre of land? The answer is $600, because that is all it will add to profit. In this problem, each of the 175 acres of land added $600 to total maximized profit of Z = $115,000. The dual solution also works in reverse. If one had to cut back on acres of land by one acre, how would it impact Z? Simply, one would lose $600 for that acre or for as many acres as would be cut back. So, to cut back 10 acres (reducing b_1 from 175 to 165), what would the impact on Z be? It would result in a $6,000 reduction ($600 × 10 acres). There are limitations to the number of units that can increase or decrease a right side value such that the dual price remains true. Those will be addressed by sensitivity analysis later.

Continuing the farming problem, the dual price values of 0 and 0 mean that the marginal contribution of an hour of labor or an extra pig are both $0. So, what is a good dual price if one was to buy an additional hour of labor or breed an extra pig beyond the 250 unit maximum limit? The answer is $0, because neither will add to Z. In the farming problem, the dual price value of 50 means that the marginal contribution of breeding an extra sheep beyond the maximum limit of 200 is $50. How can one breed an extra sheep when all 175 acres of land have been used? Well, there's an economic trade-off. To raise an extra sheep, one would have to raise one fewer pig, because there are only 175 acres for breeding. It is known that it takes 0.5 acres of land to breed a pig or a sheep. So, shifting 0.5 acres to breed an extra sheep results in losing the 0.5 acres from a pig. The result is gaining $350 for a new sheep but losing $300 for the pig, or a net economic trade-off of $50 ($350 – $300). Note that it is only in this simple problem that one can easily illustrate the trade-off. The dual solution also works in reverse. Cutting back on sheep breeding by one, how would this impact Z? Simply, one would lose $50 for that sheep and for as many additional sheep as would be cut back. Finally, in the farming problem, the dual Reduced Cost values of 0 and 0 mean that the farmer will incur $0 marginal cost or loss of breeding of the 150 pigs and 200 sheep.

The sensitivity analysis information is available in LINGO. To aid in understanding the LINGO printouts, summary tables are reprinted from LINGO in this appendix. The columns (Allowable Increase and Allowable Decrease) in the following Decision Variables table define the sensitivity analysis boundaries for the c_j parameters. From the farming problem printout, one can determine the c_j sensitivity analysis relevant ranges as follows:

Decision Variable	Current c_j Coefficient	Allowable Increase	Allowable Decrease	Relevant Range
X_1 (Pigs)	300	50	300	0 to 350
X_2 (Sheep)	350	1E + 30	50	300 to No Limit

The LINGO value of 1E + 30 is to be viewed as a very large or no-limit number. So, in this problem, one can answer the following types of decision-making questions:

1. What is the lowest profit level on a pig, and will they still be bred? (Answer: $0)
2. What is the lowest profit level on a sheep, and will they still be bred? (Answer: $300)
3. If the profit on pigs increases to $400, will sheep still be bred? (Answer: No, the sheep variable will drop out of the solution.)
4. If the profit on each sheep drops from $350 to $320, will one still breed sheep? Can the new Z value be computed without rerunning the model? (Answer: Yes to both questions, since it is within the relevant range. The old Z of $115,000 will incur a decline to $109,000 [a reduction of $30 × 200 sheep]).
5. What if pig profit goes from $300 to $310? Should one breed more pigs? Will more profit be made? (Answer: No, the solution remains the same, so by not breeding more pigs, more profit will be made. The increase in profit will be $1,500 [$10 × 150 pigs]).

The following constraints table defines the boundaries for the b_i parameters. In the restructured LINGO farming problem's printout, the b_i sensitivity analysis relevant ranges can be determined as follows:

Constraint	Dual Variable	Current b_i Coefficient	Allowable Increase	Allowable Decrease	Relevant Range
1 (land/acre)	$600	175	25	75	100 to 200
2 (labor hours)	0	7000	1E+30	750	6250 to No limit
3 (pigs)	0	250	1E+30	100	150 to No limit
4 (sheep)	$40	200	150	100	100 to 350

So, in this problem, the following types of decision-making questions can be answered:

1. If one must decrease the number of acres of land, how many can be decreased and still be assured to lose only $600 per acre? (Answer: 75)

2. If one has to increase the number of acres of land, how many can be increased and still have assurance of a gain of $600 per acre? (Answer: 25)

3. If one has to decrease the number of sheep bred at maximum, how many can be decreased and still be assured to lose only $50 per sheep? (Answer: 100)

4. How many hours of labor can be decreased and not change the existing solution of 150 pigs and 200 sheep? (Answer: 750, which are the slack hours)

Finally, it should be noted that if the c_j and b_i parameter boundaries as defined by the relevant range are exceeded, the solution set variables and the solution as a whole break down. In such situations, it is advisable to simply change the parameters and simulate the impact on the solution by rerunning the model in the software.

C.3.2 A Second Primal Maximization Problem

Consider the LINGO printout of a second sample LP maximization problem presented in Figure C.3.

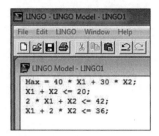

Figure C.3 Formulation of second max problem in LINGO

Define the dual price values for this problem generally as follows:

- Marginal contribution of one unit of b_1
- Marginal contribution of one unit of b_2
- Marginal contribution of one unit of b_3

Define the dual reduced cost values related to each of the primal problem's X_j decision variables as follows:

- Marginal cost or loss of one unit of X_1
- Marginal cost or loss of one unit of X_2

The LINGO printout solution for this problem is presented in Figure C.4.

```
Solution Report - LINGO1

  Global optimal solution found.
  Objective value:                        800.0000
  Total solver iterations:                        4


             Variable          Value       Reduced Cost
                   X1       20.00000           0.000000
                   X2        0.000000          10.00000

                  Row   Slack or Surplus       Dual Price
                    1          800.0000         1.000000
                    2          0.000000        40.00000
                    3          2.000000         0.000000
                    4         16.00000          0.000000
```

Figure C.4 LINGO LP solution to the second max problem with dual solution

The printouts can be restated as follows:

Primal Solution

Values	(Slack)	Related Dual Solution Values
$b_1 = 20$	$(s_1 = 0)$	40 (Marginal contribution of unit of b_1)
$b_2 = 42$	$(s_2 = 2)$	0 (Marginal contribution of unit of b_2)
$b_3 = 36$	$(s_3 = 16)$	0 (Marginal contribution of unit of b_3)
$X_1 = 20$		0 (Marginal cost or loss of one unit of X_1)
$X_2 = 0$		10 (Marginal cost or loss of one unit of X_2)

The dual price value of 40 means the marginal contribution of one unit of b_1 is $40. So, if b_1 is increased to 21, one will get an additional $40 added to Z, and if b_1 is decreased to 19, one will lose $40 from Z.

The other dual price values of 0 and 0 mean that the marginal contribution of increasing either right side value (b_1 or b_2) will not add anything to Z. This is expected, because there are slack resources in both of these constraints.

In this problem, the reduced value of 0 is expected because its related X_1 is equal to a positive value (20). The reduced cost value of 10 means that each unit of X_2 that one decides to produce will cost $10 in maximized Z. (One will lose it.) Because the model's current solution suggests not to produce X_2 one would not, but if a constraint is added to this model such that $X_2 = 1$, and then it is rerun, this model would force the solution to produce one unit of X_2. In such a case, one would also lose $10 from the maximized $800.

C.3.3 A Primal Minimization Problem

For this first minimization problem, consider a new word problem, which will be called the manufacturing company problem.

Problem Statement: A manufacturing company wants to determine how many days per month each of two plants (Plant A and Plant B) should be operated to satisfy at least the minimum market demand on three tires produced in each plant. The number of each tire produced per day and the total minimum demand is given in the following table:

	Daily Tire Production		Total Minimum
Type of Tire	Plant A	Plant B	Monthly Demand
Premium	50	60	2,500
Deluxe	80	60	3,000
Regular	100	200	7,000

If it costs the manufacturing company $2,500 per day to operate Plant A and $3,500 per day to operate Plant B, what are the optimal number of days each plant should be operated to satisfy the total minimum monthly demand on tires? The LP model formulation of this problem follows:

Minimize: $Z = 2{,}500X_1 + 3{,}500X_2$ (Cost)

subject to:

$50X_1 + 60X_2 \geq 2{,}500$ (Premium tires)

$80X_1 + 60X_2 \geq 3{,}000$ (Deluxe tires)

$100X_1 + 200X_2 \geq 7{,}000$ (Regular tires)

and

$X_1, X_2 \geq 0$

where:

X_1 = number of days to operate Plant A per month

X_2 = number of days to operate Plant B per month

The LINGO data entry of this model is presented in Figure C.5.

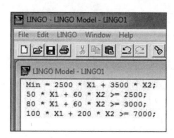

Figure C.5 Formulation of primal manufacturing company problem in LINGO

The dual problem solution values in this problem follow:

- Marginal contribution of one premium tire
- Marginal contribution of one deluxe tire
- Marginal contribution of one regular tire

These dual price values are simply the "marginal contribution of one..." unit of a right side value of a resource. The dual reduced cost values are related to each of the primal problem's X_j decision variables:

- Marginal cost or loss of day of operation of Plant A
- Marginal cost or loss of day of operation of Plant B

As stated before, it is necessary to fully understand the components of the primal problem to define the dual solution values. The LINGO printout of this solution is presented in Figure C.6.

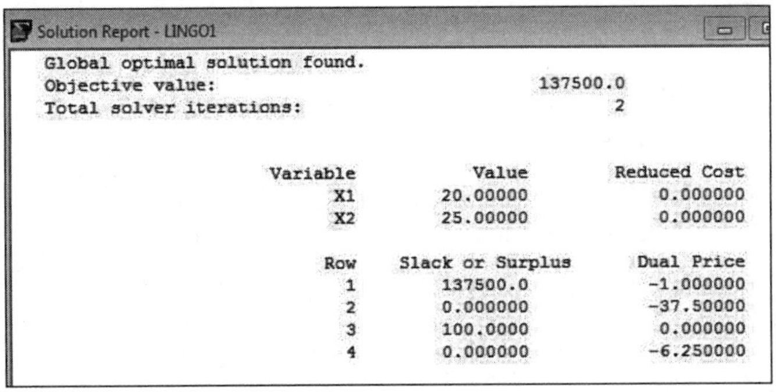

```
Solution Report - LINGO1

Global optimal solution found.
Objective value:                       137500.0
Total solver iterations:                      2


        Variable          Value        Reduced Cost
              X1        20.00000            0.000000
              X2        25.00000            0.000000

             Row    Slack or Surplus       Dual Price
               1           137500.0        -1.000000
               2           0.000000       -37.50000
               3           100.0000         0.000000
               4           0.000000        -6.250000
```

Figure C.6 LINGO LP solution to the primal manufacturing company problem with dual solution

From the printouts, the primal and dual solutions follow:

Primal Solution

Values	(Surplus)	Related Dual Solution Values
$b_1 = 2,500$	$(s_1 = 0)$	37.5 (Marginal contribution of a premium tire)
$b_2 = 3,000$	$(s_2 = 100)$	0 (Marginal contribution of a deluxe tire)
$b_3 = 7,000$	$(s_3 = 0)$	6.25 (Marginal contribution of a regular tire)
$X_1 = 20$		0 (Marginal cost or loss of one day of operating Plant A)
$X_2 = 25$		0 (Marginal cost or loss of one day of operating Plant B)

Note on the printout that the dual price values have negative signs in front of them. Ignore these signs. These denote that the value originates from greater-than or equal constraints. The interpretation of these values is quite similar to a maximization problem.

Interpreting this solution for the manufacturing company problem, the dual price value of 37.5 means that the marginal contribution of one premium tire is $37.50. So, what is a good dual price to pay for this tire? The answer is at least $37.50, because that is what it is costing in total minimized Z. The dual solution also works in reverse. If it is necessary to cut back on production by one premium tire, how would it impact Z? Simply reduce total cost by $37.50 for that one tire or on as many tires as needed to cut back. So, to cut back ten premium tires (reducing b_1 from 2,500 to 2,490), what would the impact on Z be? It would be a $375 reduction ($37.50 × 10 tires).

In the manufacturing company problem, the dual price value of 0 means that the marginal contribution of adding a deluxe tire is $0. This is expected because there is a surplus in this constraint, making it nonbinding or redundant. Adding or subtracting a deluxe tire from the 3,000 minimum value will have no impact on the existing solution.

In the manufacturing company problem, the dual price value of 6.25 means that the marginal contribution to Z of producing an extra regular tire (beyond the 7,000) is $6.25. The dual solution also works in reverse. If it is necessary to cut back on producing regular tires by one, it would reduce Z by $6.25 for that tire or on as many tires as needed to be cut back.

Finally, in the manufacturing company problem, the reduced cost values of 0 and 0 mean that the manufacturing company will incur $0 marginal cost or loss for operating Plant A 20 days per month and Plant B 25 days per month.

From the following manufacturing company problem printout, the c_j and b_i sensitivity analysis relevant ranges can be determined as follows:

Decision Variable	Current c_j Coefficient	Allowable Increase	Allowable Decrease	Relevant Range
X_1 (Plant A)	2,500	416.66	750	1,750 to 2,916.66
X_2 (Plant B)	3,500	1500	500	3,000 to 5,000

Constraint	Dual Variable	Current b_i Coefficient	Allowable Increase	Allowable Decrease	Relevant Range
1 (Premium)	$37.50	2,500	1,000	40	2,460 to 3,500
2 (Deluxe)	0	3,000	100	1E+30	No Limit to 3,100
3 (Regular)	$6.25	7,000	222.22	2,000	5,000 to 7,222.22

The interpretation of the relevant ranges is similar to the maximization problem, except Z is a cost function. Increases up to the boundaries and the dual solution trade-offs remain true. Beyond the boundaries invalidates the dual solution values. One interesting b_i relevant range value for deluxe tires indicates the lower boundary is No Limit when in fact zero has to be the lower boundary. The use by LINGO here of No Limit or 1E + 30 is simply a default. Users should be aware that zero has to be the boundary to ensure that the nonnegativity requirements of the LP model are valid.

C.3.4 A Second Primal Minimization Problem

Consider the printout of the LP maximization problem in Figure C.7.

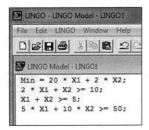

Figure C.7 Formulation of second min problem in LINGO

The dual price values for this problem can generally be defined as such:

- Marginal contribution of one unit of b_1
- Marginal contribution of one unit of b_2
- Marginal contribution of one unit of b_3

The dual reduced cost values related to each of the primal problem's X_j decision variables follow:

- Marginal cost or loss of one unit of X_1
- Marginal cost or loss of one unit of X_2

The LINGO printout of this solution is presented in Figure C.8.

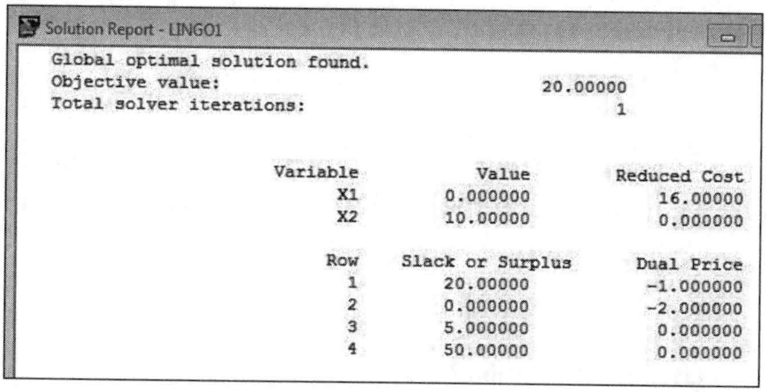

```
Solution Report - LINGO1

Global optimal solution found.
Objective value:                      20.00000
Total solver iterations:                     1


          Variable          Value       Reduced Cost
                X1       0.000000          16.00000
                X2      10.00000           0.000000

          Row    Slack or Surplus       Dual Price
            1        20.00000           -1.000000
            2         0.000000           -2.000000
            3         5.000000            0.000000
            4        50.00000            0.000000
```

Figure C.8 LINGO LP solution to the second min problem with dual solution

The LP solutions can be restated as follows:

Primal Solution		
Values	**(Surplus)**	**Related Dual Solution Values**
$b_1 = 10$	$(s_1 = 0)$	2 (Marginal contribution of unit of b_1)
$b_2 = 5$	$(s_2 = 5)$	0 (Marginal contribution of unit of b_2)
$b_3 = 50$	$(s_3 = 50)$	0 (Marginal contribution of unit of b_3)
$X_1 = 0$		16 (Marginal cost or loss of one unit of X_1)
$X_2 = 10$		0 (Marginal cost or loss of one unit of X_2)

Interpreting this solution, the dual price value of 2 (the negative sign is ignored) means that the marginal contribution of one unit of b_1 is $2. So, if b_1 is increased from 10 to 11, one will have to add an additional $2 to Z, and if b_1 is decreased to 9, Z will be decreased by $2.

The other dual price values of 0 and 0 mean the marginal contribution of increasing either right side value will not add anything to Z. This is expected, because there are surplus resources in both of these constraints.

In this problem, the reduced cost value of 16 means that for each unit of X_1, it will cost $16 in minimizing Z. (One will have to add $16 to Z.) Because the model's current solution suggests not to produce X_1, it will not be produced. However, if a constraint is added to this model such that $X_1 = 1$, and then it is rerun, the solution would be forced to produce one unit of X_1. In such a case, Z would be increased by $16, from the minimized value of $20 up to $36. The value of the other reduced cost value of 0 is expected because its related X_2 is equal to a positive value (10).

C.4 Determining the Economic Value of a Resource with Duality

The procedure for determining the economic value of a resource using duality is quite simple once the primal formulation and the dual solution are known. In the farming problem from Section C.3.1, the dual solution was given as follows:

Primal Solution

Values	(Slack)	Related Dual Solution Values
$b_1 = 175$	$(s_1 = 0)$	600 (Marginal contribution of an acre of land)
$b_2 = 7,000$	$(s_2 = 750)$	0 (Marginal contribution of an hour of labor)
$b_3 = 250$	$(s_3 = 100)$	0 (Marginal contribution of a pig)
$b_4 = 200$	$(s_3 = 0)$	50 (Marginal contribution of a sheep)
$X_1 = 150$		0 (Marginal cost or loss of a pig)
$X_2 = 200$		0 (Marginal cost or loss of a sheep)

To determine the economic contribution of each of the four resources (acres, labor hours, pigs, and sheep constraints), all that is needed is to multiply their marginal contribution coefficient by the actual number of units of each resource in the final solution. In the farming problem, this would be in order of each constraint:

Total contribution of acres of land = 175 × $600 = $105,000

Total contribution of hours of labor = 6250 × $0 = 0

Total contribution of an extra pig = 150 × $0 = 0

Total contribution of an extra sheep = 200 × $50 = 10,000

Total maximized profit (Z) = $115,000

So, acres contribute most of the $115,000 maximized profit in this problem. Note that labor hours contribute nothing. How can this be when 6,250 hours of labor are used in the resulting optimal solution? It can be so because this instance looks only at the economic value of scarcity of resources, not an accounting value based on the actual cost of breeding animals. Because labor hours are slack, they are viewed as a free economic resource that does not impact the solution. And indeed, labor hours did not in any way determine the optimal values of the decision variables in this problem.

C.5 Duality Practice Problems

Following are several practice duality problems, each followed by the answer. Use these problems to practice the methodologies and concepts presented in this appendix.

1. A small furniture manufacturer produces three different kinds of furniture: desks, chairs, and bookcases. The wooden materials have to be cut properly by machines. In total, 100 machine-hours are available for cutting. Each unit of desks, chairs, and bookcases requires 0.8 machine hours, 0.5 machine hours, and 0.5 machine hours, respectively. This manufacturer also has 650 labor hours available for painting and polishing. Each unit of desks, chairs, and bookcases requires five labor hours, three labor hours, and three labor hours for painting and polishing, respectively. These products are to be stored in a warehouse, which has a total capacity of 1,260 sq. ft. The floor space required by these three products is 9 sq. ft., 6 sq. ft., and 9 sq. ft., respectively, per unit of each product. In the market, each product is sold at a profit of $30, $16, and $25 per unit, respectively. What is the dual solution of this problem, and what is its interpretation? (Answer: First, start with the formulation of the primal problem. This problem was taken from the Practice Problems in Appendix B.)

   Let X_1, X_2, and X_3 be the number of units of desks, chairs, and bookcases to be produced, respectively. Because 100 total machine hours are available for cutting, the production of X_1, X_2, and X_3 should utilize no more than the available machine hours. Therefore, the mathematical statement of the first constraint is in the form: $0.8X_1 + 0.5X_2 + 0.5X_3 \leq 100$. Also, no more than 650 labor hours and 1,260 sq. ft. are available for painting, polishing, and storing, respectively. Therefore, these two constraints are in the form $5X_1 + 3X_2 + 3X_3 \leq 650$ and $9X_1 + 6X_2 + 9X_3 \leq 1,260$. Finally, the decision variables must be nonnegative. The complete problem formulation is as follows:

   Maximize: $Z = 30X_1 + 16X_2 + 25X_3$
   subject to: $0.8X_1 + 0.5X_2 + 0.5X_3 \leq 100$
   $5X_1 + 3X_2 + 3X_3 \leq 650$
   $9X_1 + 6X_2 + 9X_3 \leq 1,260$
   and $X_1, X_2, X_3 \qquad \geq 0$

Answering the rest of this problem requires computer usage. What are the optimal values for primal and dual solutions, and what do they mean? Answer: The primal solution is where $Z = 4,000$, $X_1 = 100$, $X_2 = 0$, $X_3 = 40$, $s_1 = 0$, $s_2 = 30$, and $s_3 = 0$; the dual solution is where the dual prices are 16.667, 0, 1.852, and the dual reduced cost values are 0, 3.444, 0, respectively. What does this dual solution mean? The marginal contribution of one machine hour is worth $16.667, labor hours are worth $0, and each square foot is worth $1.852. Also, if one decides to produce chairs (even though the primal solution says not to), it will cost $3.444 in profit per unit.

2. The Riverside Company wants to outsource the production of three products: premium zizs, deluxe zizs, and regular zizs. These three zizs can be produced at two different external plants with unique production capacities. In a normal day, Outsource Plant A produces 20 premium zizs, 30 deluxe zizs, and 100 regular zizs. Outsource Plant B produces 50 premium zizs, 50 deluxe zizs, and 60 regular zizs. The monthly demand for each of the zizs is known to be 5,000 units, 3,000 units, and 1,000 units, respectively. The company pays a daily cost of operation of $50,000 for Outsource Plant A and $50,000 for Outsource Plant B. What is the dual solution of this problem, and what does it mean?

Answer: Again, formulate the primal problem. Let the decision variables X_1 and X_2 represent the number of days of operation in each of the plants. The objective function of this problem is the sum of the daily operational costs in the two different plants in the form: Minimize: $Z = 50,000X_1 + 50,000X_2$. The objective is to determine the value of the decision variables X_1 and X_2, which yield the minimum total cost subject to the constraints. The production of each of three different zizs must be at least greater-than or equal-to the specific quantity to meet the demand requirements. In no event should the production be less than the quantities of each product demanded. Together with the constraints, the problem can be formulated as follows:

Minimize: $Z = 50,000X_1 + 50,000X_2$

subject to: $20X_1 + 50X_2 \geq 5,000$ (premium zizs)

$30X_1 + 50X_2 \geq 3,000$ (deluxe zizs)

$100X_1 + 60X_2 \geq 1,000$ (regular zizs)

and $X_1, X_2 \geq 0$

This problem requires computer usage. What are the optimal values for the primal and dual solutions? Answer: Primal solution is: $Z = 5{,}000{,}000$, $X_1 = 0$, $X_2 = 100$, $s_1 = 0$, $s_2 = 2{,}000$, $s_3 = 5{,}000$; dual price values 1,000, 0, 0, and dual reduced cost values, 30,000 and 0, respectively. What does this dual solution mean? Each unit of premium zizs that must be produced will add $1,000 to costs but will add $0 to costs if one produces additional deluxe or regular zizs. Also, if the decision is to use Outsource Plant A, it will add $30,000 per day to the total costs Z.

3. **(This problem requires computer software.)** What are the relevant ranges for the c_j and b_i parameters from the printout in Figure C.7? Answer:

Decision Variable	Current c_j Coefficient	Allowable Increase	Allowable Decrease	Relevant Range
X_1	20	1E + 30	16	4 to No Limit
X_2	2	8	2	0 to 10

Constraint	Dual Variable	Current b_i Coefficient	Allowable Increase	Allowable Decrease	Relevant Range
1	$2	10	1E + 30	5	5 to No Limit
2	0	5	5	1E + 30	No Limit to 10
3	0	50	50	1E + 30	No Limit to 100

D

Integer Programming

D.1 Introduction

D.1.1 What Is Integer Programming?

In the prior chapters on linear programming (LP), the values of the decision variables were allowed to be any real number, which can include fractional or decimal values. Integer programming (IP), which can also be called *integer linear programming*, is a special case of LP in which the values of the n (number of) decision variables must be integers (0, 1, 2, and so on). This means that the formulation of the IP problem/model differs from the regular LP problem/model only in regard to the statement of the given requirements of the resulting solution. That is, we change the nonnegativity and given requirements from a set of real numbers:

and $\quad X_1, X_2, \ldots, X_n \geq 0$

to the all integer programming problem/model form:

and $\quad X_1, X_2, \ldots, X_n \geq 0$ and all integer

It is possible to solve for a set of decision variable values that include both integer and noninteger (or real) values. This type of solution is called a *mixed integer programming* problem/model. In the mixed integer programming problem/model formulation, one would designate which decision variables will be integer and which will not. Consider a four-decision variable problem in which real (noninteger) values are needed for decision variables X_1 and X_2 and integer values are needed for decision variables X_3 and X_4. The nonnegativity and given requirements in such a mixed integer IP model would be as follows:

and $X_1, X_2 \geq 0; X_3, X_4 \geq 0$ and all integer

Many business analytic problems require this additional integer solution. To optimize modeled problems dealing with assigning people or producing whole units of a product, new product selection, or project selection decisions, choose IP over LP. IP has application in the third step in the BA process: *prescriptive analytics*.

D.1.2 Zero-One IP Problems/Models

Some IP problems require all integer solutions and some mixed integer solutions. Decision variables that must be integer can range from 0, 1, 2, ..., up to any integer number. That is one type of IP problem/model. In addition, there are other even more specialized IP models. One is called the *zero-one programming* (ZOP) model, which restricts the decision variable values to integer values of either zero or one. This changes the nonnegativity and given requirements to this:

and $X_1, X_2, \ldots, X_n = 0$ or 1

One might think this limitation is so restrictive that there cannot be much use for this model. Yet most day-to-day decision-making involves either a yes or a no. In most ZOP models, the decision variables are similarly used to model this decision as follows:

$X_j = 1$ (a "yes" decision)

$X_j = 0$ (a "no" decision)

D.2 Solving IP Problems/Models

D.2.1 Introduction

The *branch-and-bound method* is a solution procedure that can solve any IP problem for either all integer or mixed integer solutions using the regular simplex software. LINGO (from Lindo Systems, www.lindo.com) is based on the branch-and-bound method (that is, creating branching constraints to narrow the number of solutions to an IP or ZOP solution set) to run IP and ZOP problems and is used in this appendix.

D.2.2 A Maximization IP Problem

Consider the following IP problem, which can be solved with the all integer solution using the branch-and-bound method:

Maximize: $Z = 300X_1 + 200X_2$

subject to: $5X_1 + 2X_2 \le 180$

$3X_1 + 3X_2 \le 141$

and $X_1, X_2 \ge 0$ and all integer

LINGO requires integer variables to be identified with additional notation in the model. For all integer variables (and this permits mixed integer opportunities for noninteger variables), an additional designation is given in the expression here:

@GIN (*variable name*);

Using LINGO to solve the maximization problem, the LINGO data entry information is presented in Figure D.1.

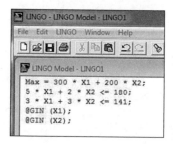

Figure D.1 A max IP problem's data entry for LINGO

The IP solution for this problem is presented in Figure D.2.

```
Solution Report - LINGO1
  Global optimal solution found.
  Objective value:                      12200.00
  Extended solver steps:                      0
  Total solver iterations:                    3


            Variable          Value       Reduced Cost
                 X1        28.00000         -300.0000
                 X2        19.00000         -200.0000

                 Row   Slack or Surplus     Dual Price
                   1        12200.00          1.000000
                   2        2.000000          0.000000
                   3        0.000000          0.000000
```

Figure D.2 A max IP solution from LINGO

D.2.3 A Minimization IP Problem

Consider the following IP problem that can be solved for the all integer solution using LINGO:

Minimize: $Z = 50X_1 + 44X_2$

subject to: $3X_1 + 2X_2 \geq 48$

$3X_1 + 6X_2 \geq 75$

and $X_1, X_2 \geq 0$ and all integer

Using this model, one can enter the problem in LINGO as presented in Figure D.3 and obtain the solution in the printout in Figure D.4.

```
LINGO - LINGO Model - LINGO1
File  Edit  LINGO  Window  Help

LINGO Model - LINGO1
 Min = 50 * X1 + 44 * X2;
 3 * X1 + 2 * X2 >= 48;
 3 * X1 + 6 * X2 >= 75;
 @GIN (X1);
 @GIN (X2);
```

Figure D.3 A min IP problem data entry for LINGO

```
Solution Report - LINGO1
 Global optimal solution found.
 Objective value:                       896.0000
 Extended solver steps:                        0
 Total solver iterations:                      4


              Variable          Value        Reduced Cost
                    X1       10.00000           50.00000
                    X2       9.000000           44.00000

                   Row    Slack or Surplus        Dual Price
                     1       896.0000           -1.000000
                     2       0.000000            0.000000
                     3       9.000000            0.000000
```

Figure D.4 A min IP solution from LINGO

To allow one of the decision variables to be a real number to achieve a mixed integer solution, all that is needed is to remove the @GIN statement in the data entry for the model.

D.3 Solving Zero-One Programming Problems/ Models

As previously stated, a ZOP model requires the values of its decision variables to be either zero or one. In the literature, there are enumeration methods (that is, methods that enumerate all possible solutions) used to solve ZOP model/problems, but these methods are beyond the scope of this course. LINGO and Excel are used to solve zero-one programming problems. To illustrate the solution process, consider a ZOP maximization model like this:

Maximize: $Z = 23X_1 + 31X_2$

subject to: $X_1 + 4X_2 \leq 34.5$

$3X_1 + 2.5X_2 \leq 45.5$

and $X_1, X_2 = 0$ or 1

LINGO requires zero-one variables to be identified with additional notation in the model. For all zero-one variables (and this permits mixed zero-one variable capabilities), an additional designation is given in the expression here:

@BIN (*variable name*);

Using LINGO to solve the maximization problem, the LINGO data entry information is presented in Figure D.5, and the solution is presented in Figure D.6.

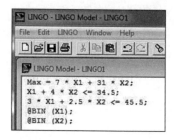

Figure D.5 A ZOP problem data entry for LINGO

```
Solution Report - LINGO1
Global optimal solution found.
Objective value:                              38.00000
Extended solver steps:                               0
Total solver iterations:                             0

                 Variable           Value        Reduced Cost
                       X1        1.000000           -7.000000
                       X2        1.000000           -31.00000

                      Row    Slack or Surplus      Dual Price
                        1        38.00000            1.000000
                        2        29.50000            0.000000
                        3        40.00000            0.000000
```

Figure D.6 A ZOP solution from LINGO

D.4 Integer Programming Practice Problems

What follows are three practice problems, followed by their answers. Use these problems to practice the methodologies and concepts presented in this appendix.

1. (Answer requires use of computer.) What is the solution to this IP problem/model?

Maximize: $Z = 4X_1 + 3X_2$

subject to: $1\,X_1 + 2\,X_2 \le 100$

$8\,X_1 + 5\,X_2 \le 300$

and $X_1, X_2 \ge 0$ and all integer (Answer: $Z = 172, X_1 = 10, X_2 = 44$)

2. (Answer requires use of computer.) What is the solution to the IP problem/model that follows?

Maximize: $Z = 40\,X_1 + 30\,X_2$

subject to: $2\,X_1 + 2\,X_2 \le 59$

$3\,X_1 + X_2 \le 75$

$X_1 + 2\,X_2 \le 50$

and $X_1, X_2 \ge 0$ and all integer (Answer: $Z = 1{,}100, X_1 = 23, X_2 = 6$)

3. (Answer requires use of computer.) What is the solution to this zero-one problem/model?

Minimize: $Z = 2X_1 + X_2 + 3X_3 + 2X_4 + 4X_5$

subject to: $X_1 + 2X_2 + X_3 + X_4 + 2X_5 \geq 4$

$7X_1 + X_2 - 3X_4 + 3X_5 \geq 6$

$-9X_1 + 9X_2 + 6X_3 - 3X_5 \geq -3$

and $X_1, X_2 \geq 0$ and all integer (Answer: $Z = 6$, $X_1 = 1$, $X_2 = 1$, $X_3 = 1$, $X_4 = 0$, $X_5 = 0$)

E

Forecasting

E.1 Introduction

From the book *Alice's Adventures in Wonderland*, there is an exchange between the Cheshire Cat and Alice. Alice asks, "Would you tell me, please, which way I ought to walk from here?" "That depends a good deal on where you want to get to," said the Cat. "I don't much care where," said Alice. "Then it doesn't matter which way you walk," said the Cat.

Business analytics (BA) helps managers learn of opportunities and solutions to problems. Making BA work requires predicting the future—or at least trends into the future. Statistical and quantitative methodologies can be used to explore and aid in understanding basic relationships within data sets. These methods are at the heart of the second step in the BA process: *predictive analytics*.

The purpose of this appendix is to introduce forecasting methodologies that are useful in exploring, conceptualizing, and predicting relationships within data. This appendix begins with a discussion of the types of variation that can be found in data and then presents a number of forecasting methodologies. The approach here will be to use SAS software to perform the computations for the analytics.

E.2 Types of Variation in Time Series Data

Forecasting in business is usually time related. When business data is matched to periods, it is called *time series data*. Time series data can be used in a forecasting model to project future sales or product demand, where time is the *predictive variable* (also called the *independent variable*) represented in the appendix by the letter X. One might want to predict sales or product demand. In a time series model, the sales or product demand data is called the *dependent variable* and is represented in the forecasting models as the letter Y.

Time series data can contain numerous unique variations that increase the complexity in forecasting and the resulting model error in predicting Y. This complexity is chiefly due to the types of variation that exist in the time series data. Four common types of variation can be present in time series data: trend, seasonal variation, cyclical variation, and random variation.

These types of variation are presented graphically in Figure E.1. One or more of them are present in time series data. For some companies, sales are dominated by a single type of variation. As such, these firms need forecasting methodologies that can accurately target the long-term (longer than one year) or short-term (less than one year) nature of these variations, as well as their linearity and nonlinearity.

TREND VARIATION

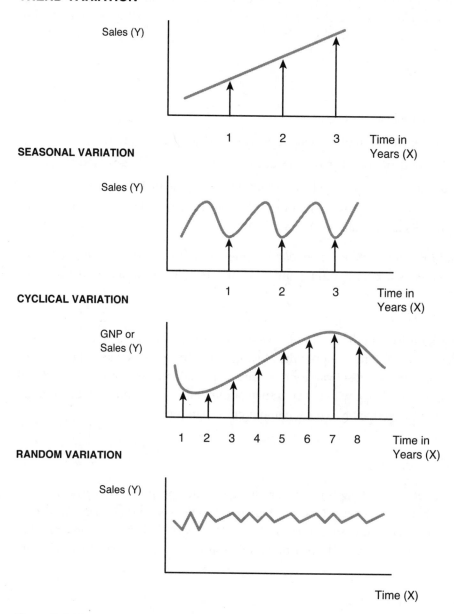

Figure E.1 Types of variation in times series data

There are two basic types of time series analysis models: an additive model and a multiplicative model. The *additive time series model* assumes that actual data is represented by an additive function, where:

$$Y_t = T_t + S_t + C_t + R_t$$

and where:

Y_t = the actual value in period t

T_t = the contribution of secular trend to the value of Y in period t

S_t = the contribution of seasonal variation to the value of Y in period t

C_t = the contribution of cyclical variation to the value of Y in period t

R_t = the random or residual contribution not explained by the other variance components in the value of Y in period t

The *multiplicative time series model* assumes that the contribution of each variance component is a compounding function of interrelated variance components, such that

$$Y_t = (T_t)(S_t)(C_t)(R_t)$$

Both the additive and the multiplicative models assume the presence, to a greater or lesser degree, of all four components of variation in time series data. The complex nature of these components of variation in time series data makes forecasting difficult in some situations. To cope with the presence of the multiple components of variation in the data, identify and analyze each component separately.

E.2.1 Trend Variation

Trend variation is a long-term linear change in an upward or downward movement of the data. Time series data that has this variation can be characterized by a gradual increasing or decreasing function over a long period of time, as presented in Figure E.1. A long-term linear forecasting method is needed to forecast this variation.

E.2.2 Seasonal Variation

Seasonal variation consists of the short-term cyclical highs and lows of behavior during a period of a year or a season that repeats itself year to year. The product sales of recreation and tourism have definite periods of high and low activity during a year. As presented in Figure E.1, this variation is nonlinear and requires a short-term, nonlinear forecasting methodology.

E.2.3 Cyclical Variation

Cyclical variation is a long-term version of seasonal variation. Cyclical variation is often described as the boom and bust periods for the economy of a country or an industry's sales. As seen by the dividing arrows in Figure E.1, there is a typical four-period sequence that an economy cyclically goes through over a long period of time from a depression period (Years 1 to 2), to a recovery period (Years 3 to 5), to a prosperity period (Years 6 to 7), to a recession period (Years 8 or more). Businesses experience similar periods of sales that may or may not be related to general economic conditions.

E.2.4 Random Variation

Random variation (see Figure E.1) is the unexplained variation that remains (the residual variation) in time series data after the other types of variation (trend, seasonal, and cyclical) have been removed. All time series data has some kind of random variation. The more random variation that's present in time series data, the more difficult it is to forecast. Indeed, if random variation is the most dominant type of variation in time series data (dominant over trend, seasonal, and cyclical variation), forecasting would nearly be impossible.

E.2.5 Forecasting Methods

Time series data can have all four types of time series variation or only one or two components. Determining the impact of these different types of variation in forecasting requires the use of several forecasting methodologies. Some methodologies are based on linear methods (for trend variation determination), and other methodologies use nonlinear methods (for seasonal and cyclical variation determination). These forecasting methodologies can identify the different types of variation and allow you to make forecasts.

Forecasting models are most useful when time series data has little or no random variation. Unfortunately, data does not always oblige with nicely linear or nonlinear data. If a pattern of variation cannot be easily discerned from time series data, then a more complex forecasting model should be employed. Some of these models will be discussed briefly in the sections that follow. If a trend, seasonal, or cyclical variation pattern in the time series data can be discerned, one might want to use one of several forecasting methods, including a simple regression model (to forecast linear trend), a multiple regression model (to forecast linear trend), or an exponential smoothing model (to forecast nonlinear seasonal or cyclical variation). In addition, other software modeling techniques can be employed to fit a model to actual data in such a way that it is useful in forecasting.

E.3 Simple Regression Model

This simple model is ideal for forecasting purposes. This section discusses how this model can be used to determine trends, how to generate a computer-based revision of this model, and where it is limited in forecasting.

E.3.1 Model for Trend

Regression (the process) allows the creation of a linear model that can be used to express a linear trend and can be used for short- or long-term forecasting. Basically, this mathematical process averages the data points to a linear expression by minimizing the distance between the data points and the line. The result is a linear model.

A *simple regression model* (one independent variable model is considered a simple model) can be employed to project a trend. The simple regression model seeks to regress data points (*X* and related *Y* points) to a single, linear expression. In a simple regression model, convert raw sales data (the *Y* variable) over time (the *X* variable) into a linear equality such that:

$$Y_p = a + bX$$
where:

Y_p = the forecasted or predicted value of the dependent variable of trend

a = vertical axis intercept value

b = slope of the trend line (denoting direction and rate of trend)

X = the independent variable, usually time in years or units of time for trend

The model parameter b provides the direction and rate of trend for each period of time (X) used in the model. If the b is positive, Y is positively related to time or X. (As X goes up in value, Y will go up.) If the b has a negative sign, Y is negatively related to X. (As X goes up, Y goes down.) Because there will be no manual computing of the values of a and b in the simple regression model shown earlier, the formulas will be omitted.

To use the model for predicting trend, the b slope value can be observed as being either positive (a positive, increasing trend into the future) or negative (a negative, decreasing trend into the future). Using the model for forecasting only requires selecting a period in the future (X) and plugging it into the model to generate a forecast value (Y_p) for that particular time.

E.3.2 Computer-Based Solution

Assume a company wants to develop a model that will reveal its sales trend and be useful in forecasting or predicting sales. For now, we will limit discussion to just the predictive variable, Time, and dependent variable, Sales. Let's assume that there are 20 months of sequentially related actual sales in the SAS data input file shown in Figure E.2. This data is presented in a more orderly fashion in Figure E.3, on which we can develop a model.

```
⊟DATA sales_data;
   INPUT time sales sun_spots ;
 CARDS;
 1 13444 78
 2 12369 20
 3 15322 1
 4 13965 83
 5 14999 5
 6 15234 2
 7 12999 1
 8 15991 4
 9 16121 21
 10 18654 5
 11 16876 45
 12 17522 5
 13 17933 17
 14 15233 8
 15 18723 2
 16 13855 12
 17 19399 5
 18 16854 1
 19 20167 24
 20 18654 4
 ;
```

Figure E.2 SAS program statement and data input file

The data from Figure E.3 can be entered using the Regression option of SAS's PROC REG DATA in Figure E.4.

The resulting simple regression model function, along with other useful statistics, is presented in Figure E.5.

	time	sales	sun_spots
1	1	13444	78
2	2	12369	20
3	3	15322	1
4	4	13965	83
5	5	14999	5
6	6	15234	2
7	7	12999	1
8	8	15991	4
9	9	16121	21
10	10	18654	5
11	11	16876	45
12	12	17522	5
13	13	17933	17
14	14	15233	8
15	15	18723	2
16	16	13855	12
17	17	19399	5
18	18	16854	1
19	19	20167	24
20	20	18654	4

Figure E.3 SAS sales and other data for forecasting model development

```
proc reg data= sales_data;
model sales=time;
output out=d1 predicted=py1;
run;
quit;
```

Figure E.4 SAS's PROC REG DATA program function statement

```
                        The REG Procedure
                          Model: MODEL1
                     Dependent Variable: sales

              Number of Observations Read        20
              Number of Observations Used        20

                        Analysis of Variance

                                  Sum of        Mean
     Source           DF         Squares       Square    F Value   Pr > F

     Model             1        51959190     51959190     20.69    0.0002
     Error            18        45196196      2510900
     Corrected Total  19        97155386


              Root MSE            1584.58189   R-Square    0.5348
              Dependent Mean         16216     Adj R-Sq    0.5090
              Coeff Var            9.77190

                        Parameter Estimates

                          Parameter     Standard
     Variable      DF      Estimate        Error     t Value   Pr > |t|

     Intercept      1         13281     736.08813      18.04    <.0001
     time           1     279.52481      61.44745       4.55    0.0002
```

Figure E.5 SAS simple regression model statistics

Additionally, charts can be provided to conceptualize the spread of actual data around the predicted linear regression model, as shown in Figure E.6 along with the SAS program statements.

```
proc gplot data=d1 ;
    goptions reset=all;
    plot sales*time py1*time/overlay legend=legend;
    symbol1 color=black v='dot';
    symbol2 color=black l=1 i=spline;
    title 'Fit of Estimated Response Function Model1';
run;
```

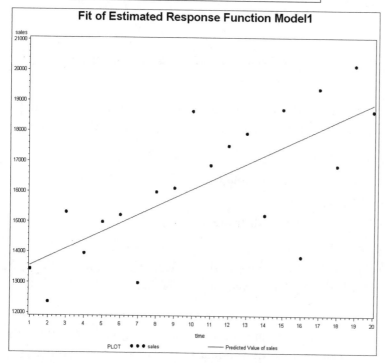

Figure E.6 SAS chart of regression model and actual sales data

Based on the SAS printout in E.5, this sales problem simple regression model can be read as follows:

$$Y_p = 13281 + 279.52581X$$

where:

Y_p = the forecasted or predicted sales value for whatever period (for example, one month)

X = any period (any month) in the future from which a forecast is desired

Because the value of b is positive (279.52581), the slope is positive, and that indicates an increasing function for sales into the future. Plugging the monthly period values of 21 and 25 into the simple regression model can generate forecast values that predict sales into the future for these two periods. So, one can predict sales of $19,151.042 (13,281+ 5,870.042) in the twenty-first period using this model, or one can look further into the future and predict sales of $20,269.145 (13281+ 6988.1452) in the twenty-fifth period. These are estimated average values, not exact ones, because it is known from the variation in Figure E.5 that the simple regression line only approximates the possible trend into the future.

E.3.3 Interpreting the Computer-Based Solution and Forecasting Statistics

To use simple regression in trend analysis and forecasting, there are some assumptions that must hold true. Some of these assumptions or rules include the following:

- There is a causal relationship between the variables X and Y.
- For every value of X, there is a distribution of Y's that allows for regressing the value of Y for predictive purposes.
- The distributions of Y are normally distributed.
- The relationship is linear.
- As the values of X fall outside the range of the value of X's used to develop the model, the accuracy of the model will increasingly be in error.

To support the use of the information from the simple regression model, statistical values and tests can aid our interpretation of the information. Looking at the SAS printout in Figure E.5, one can see that the R-Square correlation is both positive and high. (.5348 is closer to 1 than 0.) This means that as the variable Time increases, so does the forecast value for the dependent variable Sales. Also, the Sig. (1-tailed) test with a $p = 0.0002$ confirms that the relationship between Time and Sales is statistically significant.

In the Model Summary table, this correlation is further confirmed with the Adjusted R-Square statistic. (See Appendix A, "Statistical Tools," and Chapter 5, "What Is Descriptive Analytics?" for additional information on the statistical testing mentioned here.) The ANOVA F-test (actual values found in the ANOVA table) compares the linear line values with the actual values (measures the variation from the actual values to the regressed linear line). The Sig. F Change column in the Modeling

Summary and the Sig. column in the ANOVA table imply a significant relationship between the Time and Sales variables such that Time can significantly predict Sales.

The coefficients in Figure E.5 are where the a (intercept) and b (time) coefficients of the simple regression model are located. In addition, t-tests are presented (in columns t Value and Pr>|t|) that confirm in this case that both the a and the b coefficients are statistically significant.

E.4 Multiple Regression Models

One of the powerful statistical tools used in forecasting is multiple regression. We introduce this methodology in this section, provide a simple illustrative example, and explain some of the limitations on its use.

E.4.1 Introduction

Multiple regression is used to develop a model when multiple independent variables might predict a dependent variable more accurately than the one independent variable simple regression model. It is not limited to time series data but can be used to generate time series forecasts. It is ideal for sorting through possible predictive variables and determining those that should be used and those that should not be used in a forecasting model.

The generalized model for a multiple regression model can be presented as such:

$$Y_p = a + b_1 X_1 + b_2 X_2 + ... + b_n X_n$$

where:

Y_p = the forecasted or predicted value of the dependent variables of trend

a = vertical axis intercept value

X_i = (for i = 1,2, ..., n) different independent variables

b_i = (for i=1,2, ..., n) the proportional contribution of the related independent variable to the forecast of Y_p

The selection of the n different variables comes from an extensive research of possible predictive variables. These variables can be any collection or collections of data that have an observable or assumed relationship with Y.

E.4.2 Application

To illustrate the use of multiple regression, revise Sales data in Figure E.3, and this time include the additional Sun Spots variable. As previously stated for simple regression, assume a causal relationship, but in selecting Sun Spots, it is not likely to be related to the Sales variable. Regardless, both variables can be put into the multiple regression model to develop a linear regression model. The SAS model is presented in Figure E.7.

```
                              The REG Procedure
                               Model: MODEL1
                           Dependent Variable: sales

                    Number of Observations Read          20
                    Number of Observations Used          20

                              Analysis of Variance

                                    Sum of         Mean
           Source           DF     Squares        Square    F Value   Pr > F

           Model             2    52021675      26010837       9.80   0.0015
           Error            17    45133712       2654924
           Corrected Total  19    97155386


                    Root MSE            1629.39382    R-Square    0.5354
                    Dependent Mean         16216      Adj R-Sq    0.4808
                    Coeff Var           10.04825


                              Parameter Estimates

                            Parameter     Standard
           Variable    DF    Estimate       Error     t Value   Pr > |t|

           Intercept    1       13369     951.07473     14.06    <.0001
           time         1    275.32104     68.87109      4.00    0.0009
           sun_spots    1     -2.57770     16.80241     -0.15    0.8799
```

Figure E.7 SAS multiple regression printout for sales problem

The multiple regression model coefficients can be found in the Parameter Estimate column of the SAS statistics table presented in Figure E.7. The resulting multiple regression model can be taken from the printout as

$$Y_p = 13369 + 275.32104\,X_1 - 2.57776\,X_2$$

It is important to contrast the SAS printout in Figure E.5 for the single variable model when reviewing Figure E.7 to understand the potential impact of the statistics. In Figure E.7, we can see that the t value for the Time variable is still significant ($p = 0.0009$) but the t value for Sun Spots is not significant ($p = 0.8799$) at a cut-off value of $p = 0.05$.

The correlation coefficients are presented for the model as a whole. In this sales problem, the R-Square and Adjusted R-Square statistics are statistically significant. However, the slight decreases in those statistics from the single variable model are impacted by a reduction in the Adjusted R-Square (down from .5090 to .4808) and an increase in the Std. Error of the Estimate from 1,584.58189 to 1,629.39382. More error is never a good thing in forecasting. Note also that the Sig. F Change test value is now larger than 0.0015, whereas with only the one variable model, Time was more significant at 0.0002. An increase in the variance that the variable Sun Spots brought into the model caused the difference in the F-test.

Because the coefficients in the model, a, b_1, and b_2, confirm that both a and b_1 coefficients are statistically significant and the Sun Spots b_2 coefficient is not statistically significant, it is suggested that the Sun Spots variable be excluded from the model because it will bring greater variance into the forecasting and prediction efforts.

E.4.3 Limitations on the Use of Multiple Regression Models in Forecasting Time Series Data

The use of multiple regression models in time series forecasting has limitations. The application of the model to look beyond the ranges of its independent variables (like time) can violate the model's necessary assumptions previously stated for the simple regression model. Moreover, there are numerous other mathematical conditions that make forecasting with multiple independent variables risky at best. There are lag effects between independent variables that can falsely lead researchers to assume they have a fairly accurate model by bloating the correlation coefficients, when in fact the independent variables may only be correlating between themselves, not the dependent variable they seek to forecast.

A common statistical test used in regression analysis is called the Durbin-Watson Autocorrelation Test, which tests for the lagged cause-and-effect relationship between the variables. It measures the residual errors when comparing forecast values with actual values. Ideally, there should be no autocorrelation present. If there is autocorrelation present in the model, the relationship between the variables in the model is not accurately expressed. The Durbin-Watson d-test computes a statistic *d value* (similar to a t-test or Z-test) that can range from 0 to 4. The closer the value is to 2, the less residual correlations (less autocorrelation) are assumed. The closer the value is to 0, the stronger the degree of positive correlations of the residuals. The closer the d-test statistic is to 4, the stronger the negative correlation of the residuals.

SAS can generate the statistics described earlier if programmed to do so. The Durbin-Watson d-test statistic turns out in this example to be 2.837 for the multiple regression model. That suggests a slight negative correlation of residuals but not much autocorrelation. The Durbin-Watson is just one of many tests that can be run to lend validity to the use of multiple regression.

E.5 Simple Exponential Smoothing

A common forecasting method used for nonlinear forecasting is called *simple exponential smoothing*. In this section, this method is introduced and illustrated with examples.

E.5.1 Introduction

Seasonal and cyclical variations are nonlinear functions of variation. To forecast data that is dominated by these nonlinear functions, one needs a nonlinear forecasting methodology. One such forecasting methodology is exponential smoothing.

An *exponential smoothing* model, like the name implies, smooths raw data to reveal nonlinear behavior. This smoothing is accomplished by mathematically weighting the inaccuracy of a prior forecast in an effort to generate a new forecast. In other words, exponential smoothing models allow the mathematical weighting of a prior inaccurate forecast in an effort to seek to improve and make a more accurate forecast in the future. Because the model is limited to forecasting only one period into the future, it is viewed as a short-term forecasting model and can be useful in identifying cyclical and seasonal behavior in data.

The formula for the simple exponential forecasting model follows:

$$F_t = F_{t-1} + \alpha\,(A_{t-1} - F_{t-1})$$

where:

F_t = exponential smoothed forecast value for the t period

F_{t-1} = forecast value for the $t-1$ prior period

A_{t-1} = actual value for the $t-1$ prior period

α = an alpha weight ranging from 0 to 1

The values of F_{t-1} and A_{t-1} for the first forecast value of F_1 (F_0 and A_0, which are found by F_{1-1} and A_{1-1}) are usually assumed values or an average of some prior set of data. The effect of these assumed F_0 and A_0 values will eventually be averaged out by this forecasting model, so the selection can be arbitrary if the number of forecast values n is significantly large. The number of periods t to use in a model at one time, as well as the value of α in the formula, is experimental and must be determined by selecting the best combination that minimizes forecasting error. This is usually accomplished by trial-and-error methods, where various values for the parameters are substituted into the model and results simulated, whereby a comparison can be made of accuracy statistics (see Section E.8) to find the best α to use in the formula or the number of periods to run before the desired forecast value can be assumed to be generated.

E.5.2 An Example of Exponential Smoothing

Given the Time and Sales data presented in Figure E.8, one can use exponential smoothing to reveal the nonlinear cyclical (if Time is in years) or seasonal (if Time is in months or weeks) variation in the data. Using a larger α takes a larger amount of the prior actual sales value, resulting in little smoothing. Compare the two functions in Figure E.9 with differing alpha ($\alpha = 0.1$ and $\alpha = 0.9$) parameters. Using a smaller α provides a more smoothed function of the Sales variable values and reveals two nonlinear cycles in the data rather clearly. Note also that it appears the general trend is upward for the $\alpha = 0.1$ function.

	time	sales	alpha_10	alpha_90
1	1	13444	.	.
2	2	14001	13444	13444
3	3	15322	13499.7	13945.3
4	4	16965	13681.93	15184.33
5	5	18999	14010.24	16786.93
6	6	17234	14509.11	18777.79
7	7	15999	14781.6	17388.38
8	8	13991	14903.34	16137.94
9	9	11121	14812.11	14205.69
10	10	14234	14443	11429.47
11	11	16876	14422.1	13953.55
12	12	17522	14667.49	16583.75
13	13	18933	14952.94	17428.18
14	14	20233	15350.94	18782.52
15	15	21723	15839.15	20087.95
16	16	16855	16427.54	21559.5
17	17	15399	16470.28	17325.45
18	18	14854	16363.15	15591.64
19	19	11167	16212.24	14927.76
20	20	19865	15707.71	11543.08

Figure E.8 Exponential smoothing with differing alphas

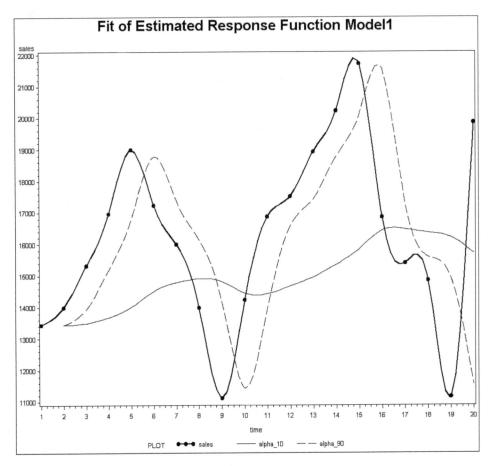

Figure E.9 Charts of exponential smoothing with differing alphas

An analyst can also use the exponential smoothing results to compute a forecast value one period out. By plugging in the last forecast value and the actual value for the twentieth period, one can derive the F_{21} forecast as follows:

$$F_t = F_{t-1} + \alpha (A_{t-1} - F_{t-1})$$
$$F_{21} = F_{20} + 0.1(A_{20} - F_{20})$$
$$F_{21} = 15707.71 + 0.1(19864 - 15707.71)$$
$$F_{21} = 16123.339$$

E.6 Smoothing Averages

Another method used to forecast nonlinear data is called smoothing averages. In this section, this method is presented with an illustrative example.

E.6.1 Introduction

A collection of averaging methods is available to forecasters to deal with the type of nonlinear data common in seasonal and cyclical variation. Some of these averaging methods include weighted moving averages. These methods seek to smooth out variations present in the data to reveal the nonlinear behavior. The forecasting model formula for a weighted moving average follows:

$$\overline{Y}_t = (w_1)Y_{t-1} + (w_2)Y_{t-2} + \dots + (w_k)Y_{t-k}$$

where:

$\overline{Y}_t$ = the forecast value in period t

Y_{t-1} = the actual value in the period just prior to period t

Y_{t-2} = the actual value of two periods prior to period t

k = the number of values to average at one time

w_i = mathematical weights such that the sum of the weights equals one

If all the w_i weights are equal, the weighted moving average becomes a moving average by simply moving one t period at a time.

Question: Given the following sales data, what is the forecast of sales for period 5 using a two-value (k) moving average with equal weights of 0.5?

Period	Sales
1	49
2	56
3	67
4	78

Answer: For the two-value average, one needs only the last two sales for periods 3 and 4:

$$\bar{Y}_t = (w_1)Y_{t-1} + (w_2)Y_{t-2}$$
$$\bar{Y}_5 = (w_1)Y_4 + (w_2)Y_3$$
$$= (0.5)78 + (0.5)67$$
$$= 72.5$$

E.6.2 An Application of Moving Average Smoothing

Once again dealing with the sales problem data now in Figure E.10, one can compute two-value moving averages and five-value moving averages (this assumes equal weighting in the smoothing model).

	time	sales	2 value	5 value
1	1	13444	.	.
2	2	12369	12906.5	.
3	3	15322	13845.5	.
4	4	13965	14643.5	.
5	5	14999	14482	14019.8
6	6	15234	15116.5	14377.8
7	7	12999	14116.5	14503.8
8	8	15991	14495	14637.6
9	9	16121	16056	15068.8
10	10	18654	17387.5	15799.8
11	11	16876	17765	16128.2
12	12	17522	17199	17032.8
13	13	17933	17727.5	17421.2
14	14	15233	16583	17243.6
15	15	18723	16978	17257.4
16	16	13855	16289	16653.2
17	17	19399	16627	17028.6
18	18	16854	18126.5	16812.8
19	19	20167	18510.5	17799.6
20	20	18654	19410.5	17785.8

Figure E.10 SAS moving average smoothing for two- and five-value averages

Note in Figure E.10 how much different the five-value moving average is from the two-value. The chart for these smoothed values is presented in Figure E.11. The five-value moving average provides the smoothest function, but it also moves the model function away from the periods when the actual sales originate. The greater the k, the greater is the movement from the actual period. This is a cost of using a smoothing methodology. Yet, it does help to recognize potential nonlinear cyclical or seasonal variations much better than just looking at the raw data. Indeed, this kind of methodology can be used in the descriptive analytics step of the business analytics process, while also having value in identifying important behavior for the predictive analytics step.

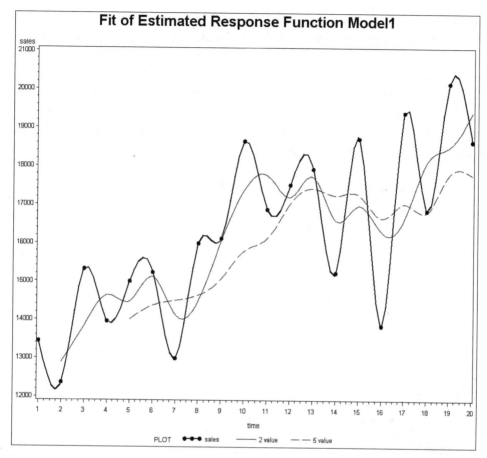

Figure E.11 SAS chart of moving average smoothing for two- and five-value averages

E.7 Fitting Models to Data

One of the many computer-based capabilities that SAS offers is a model fitting. This capability allows the analyst to take data and fit it to a predictive function using regression models.

This capability permits data of any kind, including time series data, to be fit by the software to various models in differing mathematical expressions. The process involves utilizing regression modeling to do the fitting of the data within a collection of potential models, in which each model has unique mathematical characteristics. This process permits both linear and nonlinear functions to be regressed. Also, it permits users to detect all types of time series variations and develop models to help predict them.

In Figure E.12, seven various models (linear, cubic, quadratics, and so on) are presented as they try to be fitted to a data set. In addition to fitting the data to a particular mathematical expression, this process provides statistical testing information on the model's usefulness in judging accuracy. There are many regression models available in SAS for this kind of application. Each of the seven mathematical expressions requested in this illustration used the same type of statistical information on which the best model can be selected. For information on the structure and definitions of these regression functions, see the SAS Help window within the SAS program.

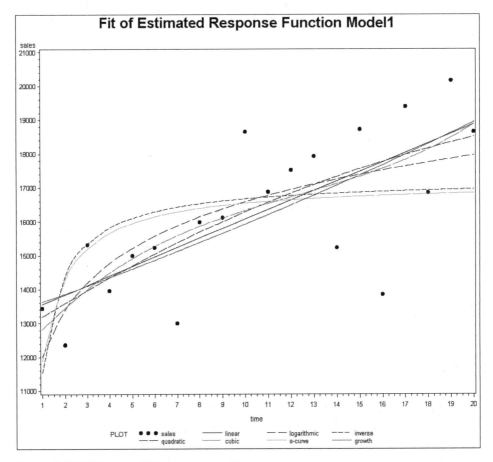

Figure E.12 SAS curve-fitting data example

In more complex models, the statistics are adjusted to that particular type of mathematical expression, such as a quadratic regression model printout included in Figure E.13. The quadratic regression model can be found from Figure E.12 to be this:

$$Y_p = a + b_1 X + b_2 X^2$$
$$= 12783 + 415.20902X - 6.46115X^2$$

Just as it was illustrated in simple regression, this model and any of the six others can be used to forecast or predict future sales. Note in the chart in Figure E.12 that none of the models includes all data points, but each can be examined in light of its t-test and F-test statistics to determine the best forecasting model.

```
                      The REG Procedure
                       Model: MODEL1
                   Dependent Variable: sales

            Number of Observations Read        20
            Number of Observations Used        20


                     Analysis of Variance

                              Sum of          Mean
    Source            DF      Squares        Square    F Value   Pr > F

    Model              2     52692092      26346046      10.07   0.0013
    Error             17     44463295       2615488
    Corrected Total   19     97155386


              Root MSE          1617.24702    R-Square    0.5423
              Dependent Mean         16216    Adj R-Sq    0.4885
              Coeff Var           9.97334


                     Parameter Estimates

                         Parameter    Standard
    Variable        DF    Estimate       Error    t Value   Pr > |t|

    Intercept        1       12783  1203.20219      10.62    <.0001
    time             1   415.20902   263.88088       1.57    0.1340
    quadratic_time   1    -6.46115    12.20572      -0.53    0.6034
```

Figure E.13 SAS quadratic regression model printout

E.8 How to Select Models and Parameters for Models

The selection of a model to forecast or a parameter in a model (for example, alpha or an independent variable in multiple regression) can be based on several criteria. The type of variation (linear or nonlinear) is one criterion. Other criteria include cost to develop a model, time it takes to develop a forecasting model, and time horizon of the forecast (long-term or short-term).

The single, most important criterion for making a final selection of a model or a parameter in a model is forecasting accuracy. Although statistical methods like correlation and t-tests provide some measure of variable relationships and their potential to predict values, in forecasting, actual results are vital. Accuracy statistics can help make this selection decision. The most accurate model will be the one that generates the least forecasting error. Several statistics (MAD, MSE, and MAPE) can be computed for any model once it has been developed. In this way, differing models can be compared, and parameters can be selected for use. Following are the formulas for these commonly used forecast accuracy statistics.

Experienced forecasters often use a simpler statistic called the *mean absolute deviation* (MAD). The formula for MAD follows:

$$\text{MAD} = \frac{\sum |A_t - F_t|}{n}$$

where:
A_t = actual value in period t
F_t = forecast value for period t
n = total number of t periods that are being summed in the numerator

The MAD statistic will be zero if the predictive model used to generate F_t perfectly predicts A_t. As the error in forecasting increases, so will the MAD statistic value. When one is comparing the MADs from different models or forecasts based on differing parameters in a model, the smaller the MAD, the more accurate is the model.

A similar statistic that seeks to minimize error in forecasting is the minimizing mean square error (MSE), using the same principles of standard error. Here's the formula for MSE:

$$\text{MSE} = \frac{\sum (A_t - F_t)^2}{n}$$

where:
A_t = actual value in period t
F_t = forecast value for period t
n = total number of t periods that are being summed in the numerator

Like the MAD statistic, the smaller the MSE, the more accurate the use of the parameter or model.

Another useful error metric is *mean absolute percentage error* (MAPE). MSE and MAPE get larger with more observations and need to be compared with other measures of the same type. MAPE has the relative advantage in that it presents errors in percentage form, making it possible to learn something about relative error immediately. Here's the formula for MAPE:

$$\text{MAPE} = \left(\frac{100}{n}\right) \sum \left|\frac{A_t - F_t}{A_t}\right|$$

E.9 Forecasting Practice Problems

Following are some practice forecasting problems, followed by the answers. Some problems can be solved by manual computation, whereas others require a computer. Use these problems to practice the methodologies and concepts presented in this appendix.

1. (Answer requires use of computer.) A company has had an annual demand of 120, 124, 127, 134, and 145 units, respectively, for the past five years. Using an alpha of 0.2, what is the forecast value for the next year? (Answer: Forecast of 6th period = 128.546)

2. (Answer requires use of computer.) A company has had an annual demand of 120, 124, 127, 134, and 145 units, respectively, for the past five years. Suppose the company wants to develop a forecasting model based on two predictive variables: Time and Index. The Time values are 1, 2, 3, 4, and 5, respectively. The Index values are 120, 135, 148, 158, and 169, respectively. What is the resulting multiple regression model? (Answer: $Y_p = 256.0505 + 21.8889X_1 - 1.3131X_2$)

3. Three models have been used to generate a forecast. Model 1's forecast has a resulting correlation coefficient of 0.79, Model 2's forecast has a resulting correlation coefficient of 0.37, and Model 3's forecast has a resulting correlation coefficient of 0.89. Which model is the best forecasting model? (Answer: Model 3 has the largest correlation coefficient. Without any other supportive statistics, it appears to be the best.)

4. Suppose sales have been calculated as a dependent variable in a regression model with Index numbers as the independent variable such that the model is $Y_p = -138.9045 + 2.0201X$. Now suppose it is found that next month's Index value is going to be X = 90. What is the predicted value of Y_p? (Answer: $Y_p = -138.9045 + 2.0201(90)$, or 42.9045.)

F

Simulation

F.1 Introduction

Mathematical models that are used to model probabilistic functions can become extremely difficult to solve. To avoid the complications and limiting assumptions of models like linear programming, simulation can be used to obtain a solution. Once a simulation model is developed and validated, it can be used to answer "what-if" questions. In the role of business analytics (BA), simulation can predict future events and payoffs. Simulations can also permit changes to systems without risk to an actual system. For example, one can assume a 5, 10, or 15 percent increase in costs in a pro forma income statement to simulate and predict the impact on profits without any risk to the organization.

F.2 Types of Simulation

Simulations can be categorized into two types: deterministic and probabilistic.

F.2.1 Deterministic Simulation

A deterministic simulation involves the use of incremental change in a parameter for a predefined model or set of equations. For example, suppose one is interested in seeing the impact on the Breakeven in Units by changing the parameter Price in the breakeven model that follows:

Breakeven in Units = Total Fixed Cost / (Price – Variable Cost)

Set the value for Total Fixed Cost equal to \$3,000 and Variable Cost equal to \$5. Vary the possible Price values to \$6, \$7, or \$8 using the preceding equation. The Price parameter changes represent the incremental change of a single parameter in the breakeven formula. The resulting simulated changes in Breakeven in Units follow:

Breakeven in Units (for Price = \$6) = 3,000 / (6 – 5) = 3,000 units

Breakeven in Units (for Price = \$7) = 3,000 / (7 – 5) = 1,500 units

Breakeven in Units (for Price = \$8) = 3,000 / (8 – 5) = 1,000 units

The three breakeven values represent deterministic simulated values. They can be used in BA to explore the impact of the three pricing scenarios.

F.2.2 Probabilistic Simulation

A probabilistic simulation occurs when we allow one or more parameters to behave in accordance with a probabilistic distribution. The *Monte Carlo simulation method* is a probabilistic simulation method. It is particularly useful in that this method does not require a specific type of probability distribution to be identified to generate a solution. Many of the advanced simulation software systems today require the identification of a specific probability distribution, like those described in Appendix A, "Statistical Tools." Many of the state-of-the-art simulators used in games and training systems are based on the Monte Carlo simulation method.

F.2.2.1 Monte Carle Simulation Method Procedure

Like all mathematical modeling approaches, the Monte Carlo simulation method requires several steps:

1. **Express system behavior as mathematical expressions, and determine the rules and assumptions under which the simulation will be run and what will determine the system's success or failure**—A mathematical expression might be a cost function. A rule might involve charging to carry stock in inventory. An assumption might be to limit the system behavior to a fixed period of time. The system's total cost might determine success or failure.

2. **Collect probability distribution information**—At least one parameter has to behave in accordance with a probability distribution. There may be dozens of parameters and probability distributions on which to collect data so that the distributions can be modeled into the simulation.

3. **Express probability distribution of each parameter in terms of a discrete distribution**—What needs to be done here is identify parameter behavior that is representative of the distributions collected. For example, sales data collected might range from $0 to $18, and one might choose to place the data into intervals such as $0 to $9 and $10 to $18, which can be called *parameter behavior*. One could then attach the observed probability to these intervals, as presented in Table F.1.

Table F.1 Paring of Parameter Behavior and Probabilities

Parameter Behavior	Probability of Behavior
1	P1
2	P2
:	:
m	Pm

The establishment of the *m* intervals allows the distribution of each parameter behavior to form a more identifiable probability distribution.

4. **Establish random number assignment system**—To establish the random number assignment system, take Table F.1 and add the numbering system to it, as presented and illustrated in Table F.2.

Table F.2 Monte Carlo Numbering System and Illustration

Random Numbering System

Parameter Behavior	Prob. of Behavior	Random Number System (Two Digits)
1	P1	00 to –
2	P2	– to –
:	:	:
m	Pm	– to 99

An Illustration

Parameter Behavior	Prob. of Behavior	Cumulative Prob. of Behavior	Random Number System (Two Digits)
$ 0 to $10	0.15	0.15	00 to 14
$11 to $20	0.20	0.35	15 to 34
$21 to $30	0.65	1.00	35 to 99

This is the heart of the Monte Carlo simulation method. The random number system can, as presented here, be a two-digit system ranging from 00 to 99. The idea is to permit a spread of 100 digits that can be proportioned to 100 percent of the Probability of Behavior column. In the example in Table F.2, there are three intervals of parameter behavior (column 1 in the table). Consider these daily sales that are observed and that one wants to simulate. In column 2 (Prob. of Behavior), a list of the observed frequency of the daily sales is expressed as a decimal. These probabilities have to add to one. In column 3 (Cumulative Prob. of Behavior), the probabilities from column 2 are added together going down from the first interval. Finally, in column 4 (Random Number System), the digits are allocated in exact proportion to the probability of the behavior starting with 00 and ending with 99. For example, there are 15 digits in the interval between 00 and 14 representing the probability of 0.15 for the Parameter Behavior interval of sales of $0 to $10. Note how the Cumulative Probability determines the upper value in the Random Number System by subtracting one from the cumulative probability (15 − 1 = 14, 35 − 1 = 34, and 100 − 1 = 99). This numbering system simulates behavior.

5. **Determine sample size for the simulation run**—Sample size can be determined in many ways. In some cases, it can be determined by time (for example, only simulate one year's worth of behavior). However, more complex statistical techniques can be used that permit statistical confidence to be included.

6. **Run the simulation, compute the desired statistics, and make decisions**— Simulations are run using computer software. In the Monte Carlo simulation method, a simulation is performed by randomly selecting a number between 00 and 99 and then determining the interval of the parameter behavior where the random number strikes. The statistics that are to be collected are usually defined in Step 1, as are the criteria on which decisions are to be based.

F.2.2.2 A Monte Carlo Simulation Application

Suppose that a company wants to determine which of two production policies should be used to set its monthly production rate. The two policies from which the company will select follow:

- **Policy 1**—Fixed monthly production rate of 100 units
- **Policy 2**—Flexible monthly production rates, in which next month's production rate is equal to last month's actual demand

The best policy is the one that will generate the least total shortage and carrying costs over a fixed period of ten months. To conduct this simulation, use the six-step Monte Carlo simulation method. Please note the following steps in this problem:

1. **Mathematical expressions, rules, and assumptions of the simulation—** The company collected the following rules, assumptions, and cost information:

   a. The number of units demanded that cannot be satisfied from monthly production is considered *inventory shortage*. A subcontractor charges $10 per unit for inventory shortage.

   b. The number of units produced in excess of demand or units carried over from the past month and that are not used in the current month are considered *carried inventory units*. Carried inventory from one month to the next month costs $2 per unit, per month.

   c. Ten months of demand will be simulated.

   d. Total Cost = Shortage Cost + Carrying Cost.

   e. Minimum Total Cost of policy over ten months determines best policy.

   f. Units carried from one month must be added to the next month's supply.

   g. Under Policy 2, the first month of production will be arbitrarily set at 100 units.

2. **Probability information—**The company has five possible demand levels. They are 80 units per month, 90 units, 100 units, 110 units, or 120 units. The respective probable occurrences observed when collecting data on the frequency of occurrences follow: 15 percent chance of 80 units of demand, 20 percent chance of 90 units of demand, 25 percent chance of 100 units of demand, 25 percent of 110 units of demand, and 15 percent chance of 120 units of demand.

3./4. **Probability distribution and random number assignment—**The table with the random number assignment schedule is presented in Table F.3.

Table F.3 Probabilities and Random Number Assignments

Parameter Behavior	Prob. of Behavior	Cumulative Prob. of Behavior	Random Number System
80	0.15	0.15	00 to 14
90	0.20	0.35	15 to 34
100	0.25	0.65	35 to 59
110	0.25	0.85	60 to 84
120	0.15	1.00	85 to 99

5. Sample Size: Given as 10 months in Step 1.

6. Simulated behavior and statistics of both policies: SAS can provide helpful statistical support to run simulations. Using SAS's SIMULATE function, one is able to load the probability distribution and data to simulate demand values that can be observed in Tables F.4 and F.5. Note in these tables that discrete probability distributions are used because the outcomes are discrete values (though we also incorporate the probability function). If desired, SAS can model the other probability distributions listed in Appendix A in place of the discrete distribution in this example.

The simulated results for Policy 1 are presented in Table F.4, and Policy 2's results are in Table F.5. Under Policy 1, a fixed production rate of 100 units per month is going to be produced to meet demand. We can see in the Production column that the 100 units are listed for all 10 months. See that the random number of 52 falls in the Random Number System interval (see Table F.3) of 35 to 59. This interval is related to the Parameter Behavior column in Table F.3 of a demand of 100 units. The first demand of 100 units has been simulated. Because Policy 1 has a fixed production rate of 100 units, and the simulated demand is 100 units, there are no Shortage Costs or Carrying Costs. This means there are 0 units, adding $0 to cost. In Month 2, a random number of 80 is drawn, which falls between the Random Number System interval of 60 to 84. That interval is associated with a demand of 110 units. Because the fixed production rate is only 100 units, the demand of 110 results in a shortage of 10 units, or a cost of $100 (10 units × $10). This process of drawing a random number, checking the interval, determining the demand, and calculating the costs is repeated for all 10 months. The resulting Total Costs for Policy 1 are $480. Now looking at the cost of Policy 2 in Table F.5, note that the demand in Month 1 is what the production rate was set at in Month 2, and the demand in Month 2 is the production in Month 3, and so on. Using the same random numbers for Policy 2 that were used for Policy 1, the resulting Total Costs are $340. Because the Total Costs for Policy 2 are less than Policy 1, select Policy 2 for the operation.

Table F.4 Resulting Simulation Policy 1 Results

Month	Random Number	Simulated Demand	Production	Unit Difference +	Unit Difference 0	Unit Difference −	Cost ($)
1	52	100	100		0		0
2	80	110	100			10	100
3	45	100	100		0		0
4	68	110	100			10	100
5	59	100	100		0		0
6	48	100	100		0		0
7	12	80	100	20			40
8	35	100	100	20			40
9	91	120	100		0		0
10	89	120	100			20	200
							Total $480

Table F.5 Resulting Simulation Policy 2 Results

Month	Random Number	Simulated Demand	Production	Unit Difference +	Unit Difference 0	Unit Difference −	Cost ($)
2	80	110	100			10	100
3	45	100	110	10			20
4	68	110	100		0		0
5	59	100	110	10			20
6	48	100	100	10			20
7	12	80	100	30			60
8	35	100	80	10			20
9	91	120	100			10	100
10	89	120	120		0		0
							Total $340

In this Monte Carlo simulation problem, only one parameter had a probability distribution. In most realistic simulation problems, many parameters are simultaneously simulated to capture the dynamics of system behavior. Modeling these types of problems requires computer software systems that specialize in simulation.

F.2.2.3 Comment on Computer Simulation Methods

Many software systems support any sized problem. The illustration here is meant only to provide a rudimentary idea of how simulation methods can work. SAS has a powerful simulation function that permits Big Data usage in both deterministic and probabilistic simulation models. The illustration of this function requires considerable programming of databases, rules, and equations, which is beyond the scope of this introductory book.

F.3 Simulation Practice Problems

Following are a couple of conceptual practice simulation problems, followed by their answers. Use these problems to practice the methodologies and concepts presented in this appendix.

1. A company has a service demand rate of 20 units 10 percent of the time, 30 units 40 percent of the time, and 40 units 50 percent of the time. Using the following random numbers (19, 45, 84, 5, 99), simulate five demand periods. Answer: 30, 30, 40, 20, and 40, respectively.

2. If there are four intervals in a simulation problem of 0 to 5, 6 to 10, 11 to 15, and 16 to 20 with related probabilities of 15 percent, 25 percent, 30 percent, and 30 percent, respectively, what "random number system" can be used to conduct the Monte Carlo simulation? Answer: 00 to 14, 15 to 39, 40 to 69, and 70 to 99.

G

Decision Theory

G.1 Introduction

Decision analysis involves a variety of methodologies that can be based on heuristics, principles, and optimization methodologies, all of which can aid decision-making. One common body of knowledge associated with decision analysis is referred to as decision theory. *Decision theory* (DT) is a field of study that applies mathematical and statistical methodologies to provide information on which decisions can be made. DT does not solve for optimal solutions like linear programming but instead is based on decision-maker preferences and principles to select choices and better satisfy needs, particularly for problem-solving environments. Before using these DT methodologies, one must know the elements of the DT model to identify and correctly formulate the problem.

The solution methodologies presented in this appendix are mathematically simple. They can easily be rendered using any spreadsheet system or SAS. Simple spreadsheets can be used to generate the prescriptive analytic information from the models presented here.

G.2 Decision Theory Model Elements

There are three primary elements in all DT problems: alternatives, states of nature, and payoffs.

1. **Decision alternatives or strategies**—The independent decision variables in the DT model that represent the alternative strategies or choices of action from which only one may be selected.

2. **States of nature**—Independent events that are assumed to occur in the future, such as an economic recession.

3. **Payoffs**—Dependent parameters that are assumed to occur if a particular alternative is selected and a particular state of nature occurs, such as improved business performance.

These three primary elements are combined into a *payoff table* to formulate the DT model. The general statement of a DT model is presented in Table G.1, where there are m alternatives and n states of nature. The idea here is that there can be a different number of alternatives than states of nature (so m does not have to equal n) and P_{ij} (where $i = 1, 2, ..., m; j = 1, 2, ..., n$) payoff values.

Table G.1 Generalized Statement of the DT Model

	States of Nature			
Alternatives	**1**	**2**	**...**	**n**
1	P_{11}	P_{12}	$...$	P_{1n}
2	P_{21}	P_{22}	$...$	P_{2n}
:	:	:	:	:
m	P_{m1}	P_{m2}	$...$	P_{mn}

G.3 Types of Decision Environments

There are three primary types of DT environments: certainty, risk, and uncertainty.

1. **Certainty**—Under this environment, the decision maker knows clearly what the alternatives are to choose from and the payoffs that each choice will bring.

2. **Risk**—Under this environment, some information on the likelihood of states of nature occurring is available but presented in a probabilistic fashion.

3. **Uncertainty**—Under this environment, no information about the likelihood of states of nature occurring is available.

G.4 Decision Theory Formulation

The procedure for formulation of a DT model consists of the following general steps:

1. Identify and list as rows the alternatives to choose from.
2. Identify and list as columns the states of nature that can occur.
3. Identify and list the payoffs in the appropriate row and column.
4. Formulate the model as a payoff table.

Using this procedure, consider the following DT problem. Suppose one wants to decide between two types of promotion efforts: A or B. The payoffs depend on the states of nature. In this problem, there are two states of nature: High Demand and Low Demand. If the selection is promotion strategy A, and one experiences a High Demand condition, the payoff will be $3 million in sales. With a Low Demand state of nature, it will result in sales equal to $1 million. If the selection is promotion strategy B, and one experiences a High Demand condition, it will result in $4 million in sales. With a Low Demand state of nature, it will result in a loss of $2 million in sales. What is the DT model formulation for this problem?

Using the four-step DT procedure, formulate this model accordingly:

1. Identify and list as rows the alternatives to choose from. There are two alternatives (A and B) and only one can be chosen.
2. Identify and list as columns the states of nature that can occur. In this problem, there are two states of nature (High Demand and Low Demand), so this results in a 2-by-2 sized payoff table.
3. Identify and list the payoffs in the appropriate row and column. The payoffs are in sales: $3, $1, $4, and $–2 million.
4. Formulate the model as a payoff table. The payoff table formulation of the complete model is presented in Table G.2.

Table G.2 DT Formulation of the Promotion Selection Problem

	States of Nature	
Alternatives	**High Demand**	**Low Demand**
Promotion A	3	1
Promotion B	4	-2

Once a DT is formulated, the payoff table can be used to analyze the payoffs and render a decision. The methodologies that are used to solve a DT problem vary by type of decision environment.

G.5 Decision-Making Under Certainty

Many different criteria can be used to aid in making decisions when the decision maker knows with certainty what the payoffs are in a given state of nature. Two of these criteria are maximax and maximin.

G.5.1 Maximax Criterion

The maximax criterion is an optimistic approach to decision-making. The maximax selection is based on the following steps:

1. Select the maximum payoff for each alternative.
2. Select the alternative with the maximum payoff of the maximum payoffs from Step 1.

To illustrate this criterion, revisit the topology problem. The solution to this problem is presented in Table G.3. As can be seen, the maximum payoffs for each of the two alternatives are $3 million and $4 million in sales, respectively. Of these, the $4 million payoff is the maximum payoff, so the max of the max is $4 million with the selection of choosing to build a Promotion B alternative.

Table G.3 Maximax Solution for DT Promotion Selection Problem

	States of Nature		Max Payoff for	Max Payoff of
Alternatives	**High Demand**	**Low Demand**	**Alternatives**	**the Max**
Promotion A	3	1	3	
Promotion B	4	-2	4	4

G.5.2 *Maximin Criterion*

The *maximin criterion* is a semi-pessimistic approach that assumes the worst state of nature is going to occur, and one should make the best of it. The maximin selection is based on the following steps:

1. Select the minimum payoff for each alternative.
2. Select the alternative with the maximum payoff of the minimum payoffs from Step 1.

To illustrate this criterion, again revisit the topology problem. The solution to this problem is presented in Table G.4. The minimum payoffs for each of the two alternatives are $1 million and $–2 million in sales, respectively. Of these, the $1 million payoff is the maximum payoff, so the max of the min is $1 million with the selection of the Promotion A alternative.

Table G.4 Maximin Solution for DT Promotion Selection Problem

| | States of Nature | | | |
Alternatives	High Demand	Low Demand	Min Payoff for Alternatives	Max Payoff of the Min
Promotion A	3	1	1	1
Promotion B	4	–2	–2	

Note the differing answers earlier to the same problem, which might cause some concern. How can one criterion suggest one alternative and another criterion suggest still another alternative? Indeed, which alternative is the best? It depends on the selection of criteria that is chosen to guide decisions. An optimist would choose a maximax approach, and a pessimist would choose the maximin approach.

G.6 Decision-Making Under Risk

Many criteria can aid in making decisions when the decision maker knows the problem faced has a risk environment where states of nature are probabilistic. In such a decision environment, both the origin of the probabilities and the criteria used to make a decision are important.

G.6.1 Origin of Probabilities

In a risk problem, probabilities are attached to each state of nature. The sum of these probabilities must add to one. In Appendix A, "Statistical Tools," a number of methodologies are presented to assess probabilities. Where probabilities come from can include objective or subjective sources. *Objective source probabilities* include experimental observation of history or using some statistical formula, such as probability distribution. When using objective methods to determine probabilities, assume the following:

1. The probability of past events will follow the same pattern in the future.
2. The probabilities are stable in the process that is being observed.
3. The sample size is adequate to represent the past behavior.

If these assumptions are not valid, an alternative way of determining probabilities involves the use of *subjective source probabilities*. This involves having experts make their best guesses at what a probability should be for the states of nature. Using this approach to probability assessment requires one to assume the experts are knowledgeable of the behavior for which they are assessing probabilities, and that their judgments are reasonably accurate.

G.6.2 Expected Value Criterion

Many criteria can be used to aid in making decisions in a risk environment. Two of these criteria are Expected Value and Expected Opportunity Loss. The expected value (EV) criterion is determined by computing a weighted estimate of payoffs for each alternative. The EV criterion is based on the following steps:

1. Attach the probabilities for each state of nature to the payoffs in each row in the payoff table.
2. Multiply the probability in decimal form by each payoff and sum by row. These values are the expected payoffs for each alternative.
3. Select the alternative with the best payoff. If the problem has profit or sales payoffs, the best payoff would be the largest expected payoff. If the problem has cost payoffs, the best payoff would be the smallest expected payoff.

To illustrate this criterion with the promotion selection problem, set the probability of High Demand at 40 percent and the probability of Low Demand at 60 percent. The probabilities attached to the states of nature change this problem into a risk-type decision environment. To compute the expected values, the probabilities in percentages are changed to decimal values and multiplied by their respective payoff values. The EVs of each alternative are presented in the last column of the payoff table in Table G.5. As can be seen, the best payoff (maximum expected profit) is with the promotion A strategy at $1.8 million.

Table G.5 Expected Value Solution for DT Promotion Selection Problem

	States of Nature		
Alternatives	**High Demand (40%)**	**Low Demand (60%)**	**Expected Values**
Promotion A	3(0.40)+	1(0.60) =	$1.8 million°
Promotion B	4(0.40)+	2(0.60) =	$0.4 million

°Best expected sales payoff.

G.6.3 *Expected Opportunity Loss Criterion*

The expected opportunity loss criterion is based on the logic of the avoidance of loss. The decision using this criterion is based on minimizing the expected opportunity loss (what one stands to lose if the best decision for each state of nature is not selected). The procedure for computing the values on which this criterion is based involves the following steps:

1. Determine the opportunity loss values in not making the best decision in each state of nature. This is accomplished by selecting the best payoff under each state of nature and subtracting all the values in that column from that particular best payoff (including itself). The result of this difference is called *opportunity loss*. The opportunity loss values can be structured into an opportunity loss table represented by the same framework as the DT payoff table.

2. Attach the probabilities to the opportunity loss values, and compute expected opportunity loss values for each alternative by summing the products of the probabilities and their respective opportunity loss values.

3. Select the alternative with the minimum expected opportunity loss value computed in Step 2.

The steps to this criterion in solving the DT promotion selection problem are presented in Tables G.6 and G.7.

1. Determine the opportunity loss values in not making the best decision in each state of nature. This is accomplished by selecting the best payoff under each state of nature and subtracting all the values in that column from that best payoff. The opportunity loss values can be structured into an opportunity loss table represented by the same framework as the DT payoff table in Table G.6. So, under the High Demand state of nature if the alternative Promotion B is selected, there will be "0" opportunity loss, since this is the best possible payoff in this state of nature. Alternatively, if Promotion A is selected, there will be an opportunity loss of $1 million in sales (i.e., $4 – $3 = $1 of loss), since with that alternative $4 million could have been made instead of just $3 million.

Table G.6 Step 1 of Expected Opportunity Loss Solution for DT Promotion Selection Problem

| | **States of Nature** | | | | | |
Alternatives	**High Demand (40%)**	**Low Demand (60%)**		**Alternatives**	**High Demand**	**Low Demand**
Promotion A	3	1		Best Payoff per State of Nature	4	1
Promotion B	4	–2				

| | **States of Nature** | |
Alternatives	**High Demand (40%)**	**Low Demand (60%)**
Promotion A	3–3=**1**	1–1=**0**
Promotion B	4–4=**0**	2–(–2)=**3**

2. Attach the probabilities to the opportunity loss values and compute expected opportunity loss values for each alternative by summing the products of the probabilities and their respective opportunity loss values.

3. Select the minimum expected opportunity loss value computed in Step 2. The minimum expected opportunity loss is with the promotion A alternative with a value of only $0.4 million.

Table G.7 Steps 2 and 3 of Expected Opportunity Loss Solution for DT Promotion Selection Problem

	States of Nature		
Alternatives	**High Demand (40%)**	**Low Demand (60%)**	**Expected Opportunity Loss**
Promotion A	1(0.40)+	0(0.60)=	$0.4 million
Promotion B	0(0.40)+	3(0.60)=	$1.8 million

G.7 Decision-Making under Uncertainty

Decision-making under uncertainty means that the decision maker has no information at all on which the state of nature will occur. Although many different criteria can be used in this environment, consider the following five: Laplace, Maximin, Maximax, Hurwicz, and Minimax.

G.7.1 Laplace Criterion

The *Laplace criterion* is based on the *Principle of Insufficient Information*. It is assumed under this principle that because no information is available on any state of nature, each is equally likely to occur. As such, one can assign an equal probability to each state of nature and then compute an expected value for each alternative. The Laplace selection is based on the following steps:

1. Attach an equal probability to each state of nature.

2. Compute an expected value for each alternative using the expected value criterion.

3. Select the alternative with the best expected value computed in Step 2.

We can again illustrate this criterion by revisiting the promotion selection problem. The solution to this problem is presented in Table G.8.

Table G.8 Laplace Solution for DT Promotion Selection Problem

Alternatives	States of Nature	
	High Demand (40%)	**Low Demand (60%)**
Promotion A	3	1
Promotion B	4	–2

1. Attach an equal probability to each state of nature. Because there are two states of nature, the probability of each is 50 percent or 0.50.

2. Compute an expected value for each alternative. The expected value computations are as follows:

   Promotion A: 3(0.50) + 1(0.50) = $2 million

   Promotion B: 4(0.50) + (–2)(0.50) = $1 million

3. Select the alternative with the best expected value computed in Step 2. The best alternative is Promotion A at $2 million in sales.

G.7.2 Maximin Criterion

The *maximin criterion* is the same as it was under certainty. The solution is the same as given before.

G.7.3 Maximax Criterion

The *maximax criterion* is the same as it was under certainty. The solution is the same as given before.

G.7.4 Hurwicz Criterion

The Hurwicz criterion uses the decision maker's subjectively weighted degree of optimism of the future. The coefficient of optimism is used for this weighting. The coefficient of optimism is on a scale from 0 to 1 and is represented by the Greek letter alpha, or α. The closer alpha is to 1, the more optimistic the decision maker is about the future. The coefficient of pessimism is $1 - \alpha$. The Hurwicz selection is based on the following steps:

1. State the value of alpha or α.
2. Determine the maximum and minimum payoffs for each alternative.

3. Multiply the coefficient of optimism (α) by the maximum payoff, multiply the coefficient of pessimism $(1 - \alpha)$ by the minimum payoff, and add these values together to derive the expected value for each alternative.

4. Select the alternative with the best expected payoff from Step 3.

To illustrate this criterion again, revisit the DT promotion selection problem. The solution to this problem is presented in Table G.9.

1. State the value of α. Let $\alpha = 0.7$. This means one is more optimistic (closer to 1).

2. Determine the maximum and minimum payoffs for each alternative.

Table G.9 Hurwicz Solution to the DT Promotion Selection Problem

	States of Nature			
Alternatives	**High Demand**	**Low Demand**	**Maximum Payoff**	**Minimum Payoff**
Promotion A	3	1	3	1
Promotion B	4	–2	4	–2

3. Multiply the coefficient of optimism (α) by the maximum payoff, multiply the coefficient of pessimism $(1 - \alpha)$ by the minimum payoff, and add these values together to derive the expected value for each alternative.

Promotion A: 3 (0.7) + 1 (1 – 0.7) = $2.4 million
Promotion B: 4 (0.7) + (–2)(1 – 0.7) = $2.2 million

4. Select the best expected payoff from Step 3. The best payoff is with the Promotion A alternative at $2.4 million in sales.

G.7.5 Minimax Criterion

The *minimax criterion* is similar to the expected opportunity loss criterion in that it is based on avoidance of loss. The decision using this criterion is based on minimizing the expected opportunity loss. The procedure for computing the values based on the minimax criterion consists of the following steps:

1. Determine the opportunity loss values in not making the best decision in each state of nature. This is accomplished by selecting the best payoff under each state of nature and subtracting all the values in that column from that particular

best payoff. The opportunity loss values can be structured into an opportunity loss table represented by the same framework as the DT payoff table.

2. Determine the maximum opportunity loss values for each alternative.

3. Select the alternative with the minimum opportunity loss value determined in Step 2.

The steps to this criterion in solving for the DT promotion selection problem are presented in Table G.10.

Table G.10 Minimax Solution of the DT Promotion Selection Problem

	States of Nature				
Alternatives	**High Demand (40%)**	**Low Demand (60%)**	**Alternatives**	**High Demand**	**Low Demand**
Promotion A	3	1	Best Payoff per State of Nature	4	1
Promotion B	4	–2			

1. Determine the opportunity loss values in not making the best decision in each state of nature. This is accomplished by selecting the best payoff under each state of nature and subtracting all the values in that column from that particular best payoff. The opportunity loss values can be structured into an opportunity loss table represented by the same framework as the DT payoff table.

	States of Nature	
Alternatives	**High Demand (40%)**	**Low Demand (60%)**
Promotion A	3–3=**1**	1–1=**0**
Promotion B	4–4=**0**	2–(–2)=**3**

2. Determine the maximum opportunity loss values for each alternative.

	States of Nature		
Alternatives	**High Demand**	**Low Demand**	**Maximum Payoff**
Promotion A	1	0	1
Promotion B	0	3	3

3. Select the minimum opportunity loss value determined in Step 2. The minimum of the maximum opportunity loss values is with the Promotion A alternative with a payoff of $1 million in sales.

G.8 Expected Value of Perfect Information

The *expected value of perfect information* (EVPI) is the difference between the expected value under a decision environment of certainty and the expected value under a decision environment of risk. In other words, if one knows exactly what state of nature will exist in the future, select the payoff maximizing action (with certainty), and then compare these optimal choices with the choices made using expected value analysis (under risk). The difference in these two values would be the EVPI. The value of EVPI is an upper boundary on what one is willing to pay for perfect information on the future states of nature.

Consider the calculation of EVPI based on expected payoffs in a risk environment. By calculating the expected profits for a personal computer (PC) rental problem (as presented in Table G.11), one would select the strategy or action of making three PCs available for customers based on the maximum expected profit of $115. The expected profit of $115 represents the best decision under risk. The calculations of the expected values, however, consider all the possible event outcomes (having only one PC available, 2 PCs available, and so on). In Table G.11, the values with an asterisk (*) are the maximum profit payoffs for each event. If it was known with certainty which of the events would occur, one could select the actions that would maximize profit.

Table G.11 Expected Value Payoffs for Each Personal Computer Rental Action

Personal Computers Available to Customers

Number of Computers Requested to Rent [(X)]	Probability of Rental Occurring [P(X)]	1	2	3	4	5
1	0.15	$60°	$20	$–20	$–60	$–100
2	0.35	60	120°	80	40	0
3	0.25	60	120	180°	140	100
4	0.15	60	120	180	240°	200
5	0.10	60	120	180	240	300°
Expected Profit		$60	$105	$115	$100	$70

°Denotes the greatest profit payoff for the particular action selected

In Table G.12, the calculations of the profit under certainty are presented. The expected profit under certainty is $162. So to make the best decision for all possible outcomes, one can expect a maximum profit of $162 in rentals per day. The difference between the maximum expected profit under certainty and the maximum expected profit under risk is EVPI, or

$$EVPI = [\text{Maximum Expected Payoff (under certainty)}] - [\text{Maximum Expected Payoff (under risk)}]$$
$$= \$162 - 115$$
$$= \$47$$

So the value of obtaining perfect information in the PC rental problem is worth $47 per day.

Table G.12 Expected Value under Best Action with Payoff Certainty

Number of Computers Requested to Rent [(X)]	Probability of Rental Occurring [P(X)]	Best Action Based on Certainty Information	Profit from Best Action ($)	Expected Value ($)
1	0.15	Make 1 computer available	60	9
2	0.35	Make 2 computers available	120	42
3	0.25	Make 3 computers available	180	45
4	0.15	Make 4 computers available	240	36
5	0.10	Make 5 computers available	300	30
		Expected Profit Under Certainty = $162		

G.9 Sequential Decisions and Decision Trees

Some decision situations require a sequence of decisions to be made. Decisions that are dependent on one another in a sequence are called *sequential decisions*. One statistical methodology that is useful in understanding a sequential decision problem formulation is a decision tree. A *decision tree* is a graphical aid that can be used to depict a sequence of decisions in a horizontal tree-like structure. The branches of the tree represent the decision paths that a decision maker may choose to take in the sequence.

Consider a survey/product introduction sequential decision problem to illustrate a sequential solution procedure using decision trees. Suppose a marketing manager must decide whether to introduce a new product. The decision tree mapping of the problem is presented in Figure G.1. Note at the top of Figure G.1 that there are two sequential decisions included in the problem. The first is deciding if a survey on customer demand is to be undertaken, and the second is deciding whether to introduce a product. In each decision, there are alternative actions that can be taken, events (some with probabilities), and payoffs.

To solve this sequential decision problem, use the backward decision method. The *backward decision method* for sequential decision problems combines the economic criterion of the maximax strategy with the expected value of the probability decision criterion discussed previously in this appendix. The backward decision method begins, as its name implies, at the payoffs at the back of the second decision. Using the maximax strategy with an environment of certainty, select the maximum payoffs for all the branches in the second decision first. The payoffs are in millions of dollars. In Figure G.2, it is apparent one would choose either the action of Introduce Product with a payoff of $6 million or the action of No Introduction with a payoff of $–2 million. So the maximax strategy would be to choose to Introduce Product with payoffs of $6 million, and the No Introduction alternative would be discarded. The double bars, or ||, indicate that the branches are not chosen and discarded. Note the maximax payoffs are brought forward to the Second Decision box in Figure G.2. In effect, this problem is being worked backward from the second decision to the first.

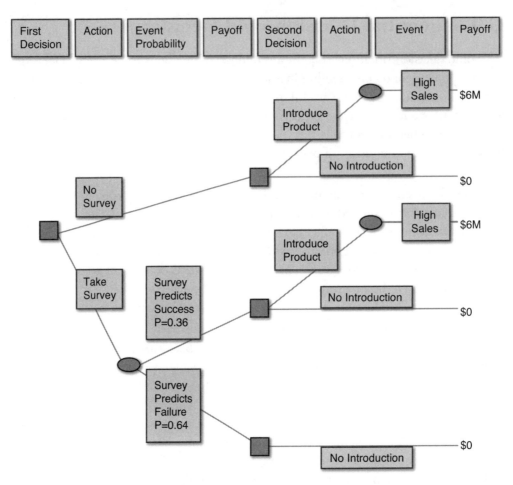

Figure G.1 Decision tree of sequential survey/product introduction problem

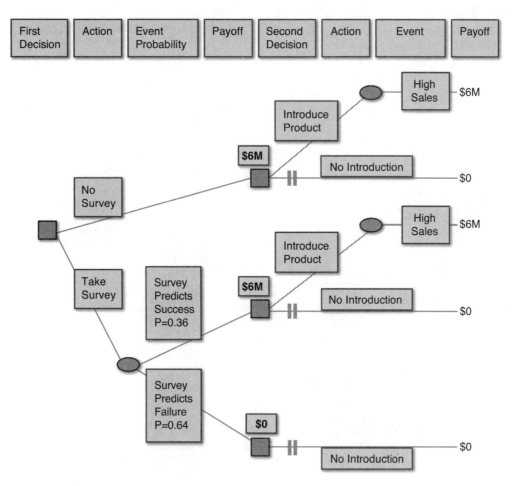

Figure G.2 Step 1 of backward solution to sequential survey/product introduction problem

We now can use the probability information on the events to calculate the expected value payoffs on which the first decision's action can be based. The expected value calculations are presented in Figure G.3. The resulting expected value of Take Survey action is $2.16 million, and the certainty payoff of No Survey is $6 million. So the manager would choose not to take a survey and introduce the product with the expectation of a $6 million payoff.

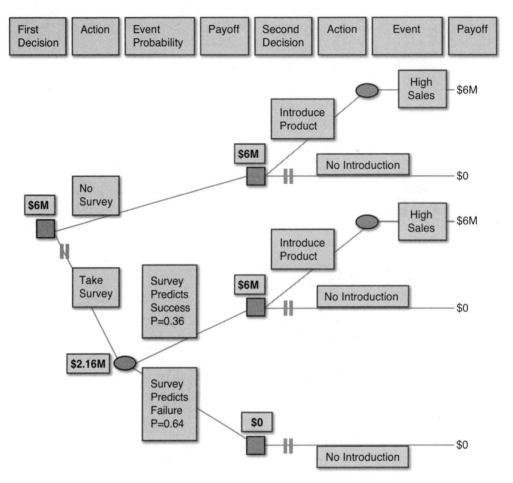

Figure G.3 Step 2 of backward solution to sequential survey/product introduction problem

This backward decision method assumes that the maximax strategy and the resulting expected values are, on average, reflective of the payoff values expected for the problem situation. The use of the solution method can be expanded to more complex problems involving three, four, or more sequential decisions. In such problems, there may be more than one set of probabilities for possible events in the sequential decision. Assume in these problems that the probabilities for each subsequent decision are statistically independent of the outcomes of each decision. If the probabilities are not independent, their conditional probability nature must be analytically considered in the sequential decision process. One way to calculate these conditional probabilities is through the use of Bayes's theorem (explained in the next section).

G.10 The Value of Imperfect Information: Bayes's Theorem

Most additional information is imperfect in that it is usually obtained in a survey or research of a sample of information, rather than a population of all information. The value of EVPI is that it provides an upper boundary of possible investment for additional information in decision-making under risk. Any information, even imperfect information that improves the chances of making a correct decision and increases the expected payoffs, may be worth the additional cost of obtaining it. The procedure by which to determine the value of imperfect additional information involves the use of Bayes's theorem.

Bayes's theorem can be used to revise prior or given probabilities by using conditional probability information (that is, the additional, imperfect information). Bayes's theorem reverses the events in a conditional probability ($P(A \mid B)$ to find $P(B \mid A)$). The formula based on Bayes's theorem that reverses conditional probabilities follows:

$$P(B|A) = \frac{P(B) \times P(A|B)}{\sum P(B_i) \times P(A|B_i)} \text{ for i=1,2,...,n}$$

where:

$P(B \mid A)$ = conditional probability of event, B, given event, A

$P(B_i)$ = probability of i = 1, 2, 3, ..., n mutually exclusive and collectively exhaustive events, B

$P(A \mid B_i)$ = conditional probability of event, A, given each event, B

Bayes's theorem is based on the rule of multiplication (see Appendix A, Section A.3.3) when events, A and B, are not independent. So the conditional probability of $P(A \text{ and } B)$ is found this way:

$P(A \text{ and } B) = P(B) \times P(B|A)$

This rule can be converted into Bayes's formula by dividing both sides of the equality by $P(B)$ or:

$$P(B|A) = \frac{P(A \text{ and } B)}{P(B)}$$

In this revised expression, the $P(B)$ denominator is called the marginal probability of all joint, $P(A \text{ and } B)$, probabilities. The *marginal probability* is the sum of the product of all $P(B_i) \times P(A \mid B_i)$. The term marginal probability comes from the fact that this probability is usually obtained from the margins (where summations of probabilities are usually located) of joint probability tables. This summation is divided into a single P(A and B) value to obtain the desired reversed conditional probability of P(B | A).

To see how Bayes's theorem is applied in business analytics and decision-making, consider a modified version of the sequential decision-making situation in Section G.9. Suppose one is facing a business decision concerning the introduction of a new product similar to the decision tree presented in Figure G.4.

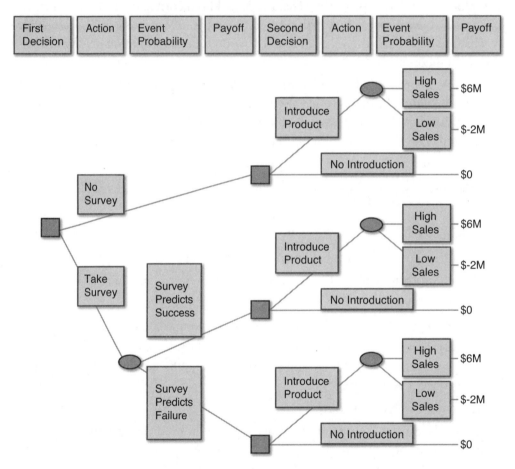

Figure G.4 Decision tree for modified survey/product introduction problem

There are two action choices: Introduce Product and No Introduction. After reviewing the new product, make a subjective judgment on the new product's sales potential. Such probabilities in Appendix A were called *subjective* probabilities because of their judgmental origin. These prior or given probabilities of the two events of High Sales and Low Sales are presented in Table G.13.

Table G.13 Payoff Table and Prior Probabilities for Survey/Product Introduction Problem

Actions		Actions	
Events	**Prior Probabilities**	**Introduce Product**	**No Introduction**
High Sales (H)	0.2	$6.0 million	$0
Low Sales (L)	0.8	$–2.0 million	$0
Total = 1.0			

Based on judgmental sales potential, the profit payoff values for each action is estimated and presented in Table G.13. Based on this prior information, we can determine the expected sales payoffs for each alternative action as:

Expected Payoff for Action:

Introduce Product = $(0.2)(\$6.0) + (0.8)(\$–2.0) = \$–.4$ million

No Introduction = $(0.2)(\$0) + (0.8)(\$0) = \$0$

Based solely on these expected sales values, choose the action of No Introduction to minimize the loss. (Losing $0 is better than losing $.4 million in sales.) On the other hand, the EVPI in this decision situation follows:

EVPI = $(0.2)(\$6.0) + (0.8)(\$0) = \$1.2$ million

The EVPI indicates that considerable expected sales are possible in this problem if one has perfect information. It can also be interpreted to justify pursuing additional imperfect information if the cost of that imperfect information is less than the expected contribution of $1.2 million that the sales will profit the organization.

This problem now becomes a sequential decision-making situation, where the first decision is whether to obtain additional information (No Survey or Take Survey), and the second decision is to Introduce Product or No Introduction. In the upper branches of the sequential decision tree in Figure G.4, the original product introduction decision is presented.

Suppose a survey is to be conducted to obtain the additional, imperfect information on which to base the product introduction decision. The purpose of the survey is to determine the successfulness of the new product. The survey will have two possible events: Survey Predicts Success or Survey Predicts Failure. The lower branches of the decision tree in Figure G.4 present this sequence of decisions. Given the problem presented in Figure G.4, one might consider using the backward solution method to determine the expected payoffs and arrive at a decision. Unfortunately, the decision cannot be made this way until one determines the probabilities for the events of survey prediction for the first decision.

To obtain the information on the events of survey prediction, one cannot use simple probabilities. The value of the probability of predicting any product success or failure may be meaningless. (It may have nothing to do with the introduction of this new product or resulting sales.) Instead, one must recognize the dependence of the event probabilities in the sequence of decisions. (Probabilities of survey results and actual sales can be related or dependent.) In this problem, decide first if additional information (via a survey) is to be obtained, and second, if the product will be introduced. Because it is necessary to start backward in the problem with the payoffs for the second decision, one must determine the probability of the second decision's events occurring, given that the first decision and its events have occurred. So, the probabilities must be determined to reflect this sequence of decision-making. Specifically, determine the conditional probabilities of the actual sales given survey results. To do this, use Bayes's theorem and some additional objective probabilistic information.

The procedure for revising prior probabilities using Bayes's theorem consists of the following steps:

1. **Obtain the prior and conditional probabilities for the events in the decision-making situation**—In the survey/product introduction problem, the prior probabilities ($P(H) = 0.2$ and $P(L) = 0.8$) are given. The conditional probabilities are the additional information that is being brought into this problem. Note based only on the prior probabilities that one would not introduce the product. The conditional probabilities can be obtained from objective sources (past history of surveys on similar products or similar surveys), or they can come from subjective sources (additional experts with experiential judgment information). In the case of the survey/product introduction problem, use the conditional probabilities presented in Table G.14. Note in that table the probability of survey results that predict a successful product, given that High Sales of 0.6 is experienced.

Table G.14 Conditional Probabilities of Survey Results, Given Actual Sales for Survey/Product Introduction Problem: P(Survey Results | Actual Results)

	Actual Level of Sales	
Survey Results	**High Sales (H)**	**Low Sales (L)**
Survey Predicts Success (S)	$P(S \mid H) = 0.6$	$P(S \mid H) = 0.3$
Survey Predicts Failure (F)	$P(F \mid H) = 0.4$	$P(F \mid H) = 0.7$
Total	1.0	1.0

2. **Convert the prior and conditional probabilities into joint and marginal probabilities**—The formulas for this conversion were presented in Appendix A and repeated for this problem in Table G.15 (A). The computations for the joint probabilities are presented in Table G.15 (B). As can be seen in Table G.15 (B), the joint probability of having High Sales and a survey result of a successful product are found by the following equation:

$$P(H \text{ and } S) = (\text{Prior probability of } H) \times (\text{Conditional probability of } S \text{ given } H)$$
$$= P(H) \times P(S \mid H)$$
$$= (0.2)(0.6)$$
$$= 0.12$$

The marginal probabilities of the survey results are found by adding the joint probabilities for all sales events. So the marginal probability of a survey predicting a product will be successful follows:

$$P(S) = P(H \text{ and } S) + P(L \text{ and } S)$$
$$= 0.12 + 0.24$$
$$= 0.36$$

Table G.15 Joint Probability Table Computations for Survey/Product Introduction Problem: P(Actual Sales and Survey Results)

A. Joint Probability Formulas for P (Actual Sales and Survey Results)

Level of Sales	Success (S)	Failure (F)	Total
High Sales (H)	P(H and S) = P(H) P(S | H)	P(H and F) = P(H) P(F | H)	P(H)
Low Sales (L)	P(L and S) = P(L) P(S | L)	P(L and F) = P(L) P(F | L)	P(L)
Total (Marginal Probabilities)	P(S)	P(F)	1.0

B. Joint Probability Computations for P (Actual Sales and Survey Results)

Level of Sales	Success (S)	Failure (F)	Total
High Sales (H)	P(H and S) = (0.2)(0.6) = 0.12	P(H and F) = (.02)(0.4) = 0.08	P(H) = 0.2
Low Sales (L)	P(L and S) = (0.8)(0.3) = 0.24	P(L and F) = (0.8)(0.7) = 0.56	P(L) = 0.8
Total (Marginal Probabilities)	P(S) = 0.36	P(F) = 0.64	1.0

3. **Compute the revised or posterior probabilities using Bayes's theorem—**
   The term *posterior probabilities* indicates that the prior probabilities have been revised to include additional probabilistic information (the conditional probabilities of survey results). Hence, the posterior probabilities are after, or posterior to, the prior probabilities. The computation of the posterior probabilities is accomplished using Bayes's theorem. The posterior probabilities for each of the end branches or payoff branches in Figure G.4 must be computed to reflect the addition of survey result information in the decision process. Using Bayes's theorem, the computation of the posterior probability of having High Sales given a survey result of the product being successful follows:

$$P(H|S) = \frac{P(H \text{ and } S)}{P(S)} = \frac{0.12}{0.36} = 0.333$$

Note that the resulting posterior probability of 0.333 is greater than the prior probability of 0.2, indicating a revision based on the additional information. The other three posterior probabilities can be similarly computed as follows:

$$P(H|F) = \frac{P(H \text{ and } F)}{P(F)} = \frac{0.08}{0.64} = 0.125$$

$$P(L|S) = \frac{P(L \text{ and } S)}{P(S)} = \frac{0.24}{0.36} = 0.667$$

$$P(L|F) = \frac{P(L \text{ and } F)}{P(F)} = \frac{0.56}{0.64} = 0.875$$

This three-step procedure can be used on any size problem where the necessary prior and conditional probabilities are available.

We can now use the marginal and posterior probabilities in combination with the backward solution method to resolve the survey/product introduction problem using sequential decision-making. In Figure G.5, the marginal and posterior probabilities are incorporated into the decision tree. As can be seen, the posterior probabilities are positioned in the end of the branches for the Take Survey decision. The expected values for each of the branches can then be computed in the same way as in the decision tree problem presented earlier. The expected values for all the No Introduction choices are $0, but now, because of the revision of the probabilities, there is a positive sales payoff of $.664 million in sales in one of the branches. Using the maximax decision criterion, choose the best payoffs from the second decision, and bring them forward to be considered in the first decision. Because the events of survey results have a probable occurrence (measured by marginal probabilities), compute a second expected value of $.239 million, representing the expected payoff for the choice of Take Survey. The decision then comes down to a choice between No Survey, which results in a $0 payoff, or Take Survey, which results in a $.239 million payoff. The best choice using the expected value criterion would be to Take Survey. If it predicts Success, one would choose to Introduce Product, and if that happens, a $.239 million payoff could be expected. If the survey predicts Failure, one would choose No Introduction, and the resulting payoff would be $0.

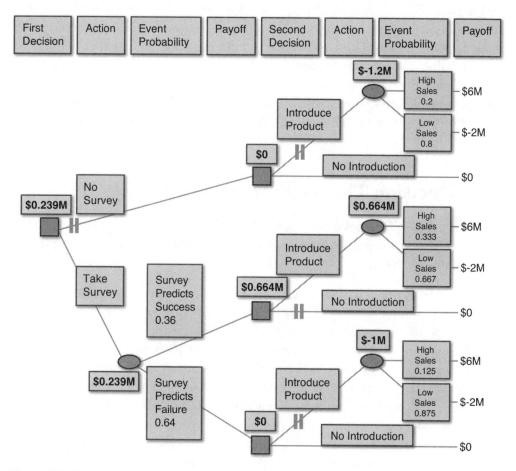

Figure G.5 Sequential decision-making solution for survey/product introduction problem

In this problem, a situation, based solely on prior probabilities, is faced in which the expected sales turned out to be $0. By adding conditional probabilities of survey results, it was determined that expected sales would increase to $.239 million. The difference occurs because of the addition of information. The information that is added is imperfect but results in an expected sales increase (over the original decision) of $.239 million.

Users should be aware of several factors when using this procedure. First, the expected value of imperfect information is an expected value and, as such, one is not assured of receiving it. Second, the additional imperfect information is yet to be obtained. Therefore, its true value is not really determinable. One can only determine if additional information is of value if it improves the ability to correctly make decisions, and if it increases the expected payoffs one expects to receive once it is used in the decision process. Third, the dependency of the event probabilities may or may not be reliable. The use of probabilities for one event to revise another event may not be valid unless the relationship can be proven. Finally, the cost of obtaining the additional information has not been discussed. If the cost to take the survey is greater than the profit in sales from the additional $.239 million, one would not take the survey.

G.11 Decision Theory Practice Problems

Below are several practice decision theory problems. Each problem is followed by the answer. Use these problems to practice the methodologies and concepts presented in this appendix.

1. A company would like to invest in one of the three types of resources: new personnel, new technology, or new processes. The firm works in an environment that permits no assurance of what the economic environment will be, nor has it information on what the environment will most likely become. The projected profit for an investment in the personnel resource can be either $2.3 million per year if a prosperous market exists or only $1.1 million if a depressed market exists. The projected profit for an investment in the technology resource can be either $2.8 million per year if a prosperous market exists or only $1.3 million if a depressed market exists. The projected profit for an investment in the process resource can be either $0.7 million per year if a prosperous market exists or $4.2 million if a depressed market exists.

a. What is the formulation of this DT model?

	State of Nature	
Alternatives	**Prosperous Market Conditions**	**Depressed Market Conditions**
Personnel	2.3	1.1
Technology	2.8	1.3
Processes	0.7	4.2

b. Which DT environment does this problem fall into: certainty or uncertainty? Answer: Because the states of nature are uncertain, this is an uncertainty problem.

c. Using the maximax criteria, what is the best choice? Answer: Processes with a payoff of $4.2 million.

d. Using the maximin criteria, what is the best choice? Answer: Technology at $1.3 million.

2. Due to a favorable stock market outcome, a firm has an opportunity to invest in new IT technologies to support new ecommerce markets they are developing. The two new markets they plan to develop sell their services to include business-to-business (B2B) and business-to-consumer (B2C). To support these markets, they can invest in one of the following three technologies: A, B, or C. The estimated yearly profit that can be provided by adding technology A in a B2B market is $3.5 million and in a B2C market is $4.0 million. The estimated yearly profit that can be provided by adding the use of technology B in a B2B market is $2.2 million and in a B2C market is $4.9 million. The estimated yearly profit that can be provided by adding the use of technology C in a B2B market is $5.0 million and in a B2C market is $2.0 million.

a. What is the formulation of this DT model?

	State of Nature	
Alternatives	**B2B**	**B2C**
Technology A	3.5	4.0
Technology B	2.2	4.9
Technology C	5.0	2.0

b. Is this problem a certainty problem or an uncertainty problem? Answer: Because the states of nature are uncertain, this is an uncertainty problem.

c. If one uses a Laplace criterion, what is the best choice? Answer: Technology A with a payoff of $3.75 million.

d. If one uses the minimax criteria, what is the best choice? Answer: Technology A with a max regret at $1.5 million.

The data that follows is used in Problems 3 and 4. Suppose one has the following cost payoff table:

	Actions		
Events	A	B	C
A	120	270	380
B	200	410	280
C	280	290	250
D	560	100	100

3. Using maximax, maximin, or a Laplace criterion, what is the best choice?

Answer: maximax – Action A (best payoff 560); maximin – Action A (worst payoff 120); Laplace – Action B (average payoff 292.5).

4. The probabilities for the four events are 0.40, 0.35, 0.15, and 0.10, respectively. What is the expected value of each of the three alternatives? Which alternative action is the best choice based on expected value analysis? Answer: A = 216, B = 305; C = 297.5; Best choice = B.

5. Recalculate the expected value of additional imperfect information for the survey/product introduction problem in Table G.13 using the revised payoff table that follows:

	Actions	
Events	Introduce Product	No Introduction
(H)	$24 Million	$0
(L)	$–12 Million	$0

Answer: EVPI = 0.2(24) + 0.8(0) = 4.8; EV(Introduce Product) = 0.2(24) + 0.8(0) = 4.8; EV(No Introduction) = 0.2(0) + 0.8(0) = 0.

6. What is the best decision in the decision trees in Figure G.6A and G.6B if one wants to maximize the expected payoffs?

Answer: B is the best choice, with an EV of 427, as opposed to 173.9 for A.

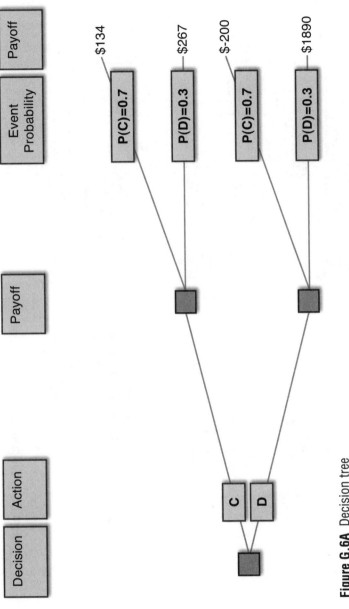

Figure G.6A Decision tree

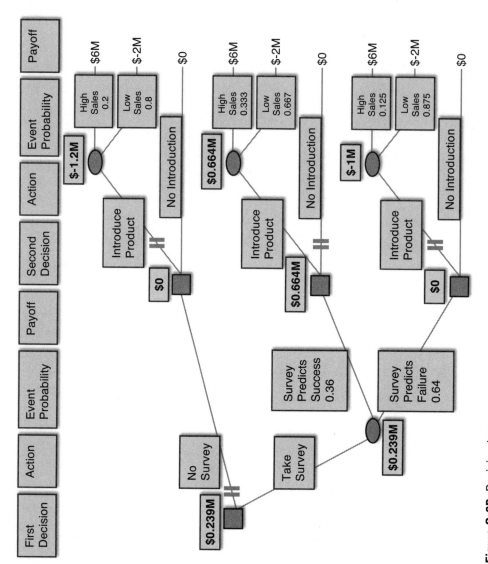

Figure G.6B Decision tree

7. Figure G.7 shows a profit maximizing decision tree. Using the backward solution method, determine the best decision. What is the EMV of the "best" decision?

Answer: No is the best choice, with an EV of 154.3.

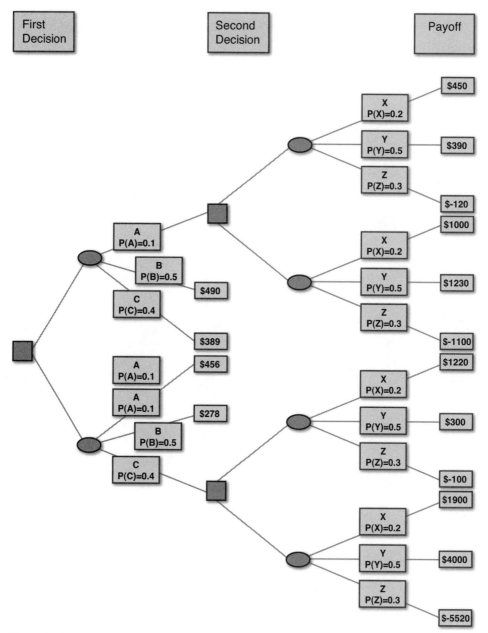

Figure G.7 Profit-maximizing decision tree

Index